BUILDING
ECONOMICS
FOR
ARCHITECTS

BUILDING
ECONOMICS
FOR
ARCHITECTS

THORBJOERN MANN

WILEY

Library of Congress Catalog Card Number 91-39713
ISBN 0-442-00389-7

Published by Van Nostrand Reinhold
115 Fifth Avenue
New York, New York 10003

Chapman and Hall
2-6 Boundary Row
London, SE1 8HN, England

Thomas Nelson Australia
102 Dodds Street
South Melbourne 3205
Victoria, Australia

Nelson Canada
1120 Birchmount Road
Scarborough, Ontario M1K 5G4, Canada

16 15 14 13 12 11 10 9 8 7 6 5 4 3 2 1

Library of Congress Cataloging-in-Publication Data

Mann, Thorbjoern, 1936–
    Building economics for architects / Thorbjoern Mann.
        p.   cm.
    ISBN 0-442-00389-7
    1. Building—Economic aspects.   I. Title.
    TH437.M36   1992
    690′.0681—dc20

                                                                91-39713
                                                                CIP

Contents

Preface

The purpose of this book is simple: to provide an introduction to those aspects of building that can help students of architecture become more aware of the economic concerns of other parties in the building process and more capable of responding constructively to these concerns in their design decisions. Carrying out this intention is not simple because the subject matter borders on many other fields of knowledge, academic disciplines, and professional domains. It is difficult to maintain a sense of balance among all these areas, and between the need to understand the principles of other disciplines well enough to engage them in a fruitful dialogue and needless trespassing onto others' turf where we architects are amateurs. It is difficult also because of a perception or preconception among some students and architects that economic concerns and architecture are somehow incompatible or even at odds with each other.

It is necessary to provide adequate coverage of all aspects of the problems without straying too far afield, to ensure understanding without becoming too theoretical, and to offer practical tools without getting lost in trivial detail. Few people are likely to agree on what the proper balance in such an undertaking should be; it is impossible to be all-inclusive. This cannot be helped. I have tried to organize the material in such a way as to lead the reader through straightforward practical concerns in a plausible sequence and to raise other interesting questions along the way, which invite further exploration and even challenge basic assumptions. Some of these explorations and excursions are taken up in the appendix, along with auxiliary and prerequisite topics that would have made for a cumbersome journey had they been placed in the main body of the book.

It is my hope that the book will help to put the economics of building more squarely on the agenda of architectural designers — not as a necessary evil, but as one more of the many concerns that make architecture an encompassing and endlessly fascinating subject. Just as adding a ball makes juggling more, not less, impressive, an additional challenge to the designer's creativity and imagination make designing an even greater art.

Acknowledgments

This book was made possible by the grant of a one-term sabbatical from the School of Architecture, Florida Agricultural and Mechanical University. I also would like to acknowledge the pervasive influence of the work of the late Professor Horst W. Rittel, from whose lectures and writings I have occasionally borrowed and whose ideas have helped me see and understand the building economics material within a larger framework of design. Most of all, I would like to thank my wife Susie for her support and encouragement.

Thorbjoern Mann

BUILDING
ECONOMICS
FOR
ARCHITECTS

1
Introduction

THE IMPORTANCE OF BUILDING ECONOMICS

The Attitude of the Architectural Profession

The attitude of architects toward issues related to the economics of building at times seems somewhat ambivalent. Most professionals agree that economic factors are quite important, in fact that they often influence design decisions more than any other single factor and should be well understood by the designer. On the other hand, in the prestigious journals where the profession represents its work to itself and the world, in discussions of architectural theory, and in the curricula of Schools of Architecture, one finds little concern for the issue. It seems tainted or even "dirty." Architects who manage to design buildings that work out well economically for their clients run the risk of being considered "hacks" by their colleagues. Even proponents of more radical alternative approaches to architecture are not immune to these attitudes, as the following incident suggests: During the question/answer period after a prestigious lecture by a well-known innovator on his new approach to building and urban design, a young builder impressed by the theory asked a question expressing his concern about how to reconcile it with the mundane economic pressures of suburban homebuilding, his means of livelihood. The speaker brushed off the questioner with the cutting reply that "we are talking about serious matters here." End of discussion. Are the economic issues of building unworthy of the architect's creative effort?

This book suggests that we take a different attitude. Not only does it urge acceptance of the economic questions of building as legitimate concerns of architectural designers; it suggests that these issues provide a relevant and worthy focus for architectural invention and creativity. It aims at providing architects with arguments they might use with clients in favor of spending more time and effort on "critical design time"—essentially, allocating more resources to careful analysis, evaluation, and comparison of design schemes at the early design stages. At the same time, it emphasizes that these arguments only will carry weight to the extent that the architect actually knows how to improve the economic performance of design solutions, and it then tries to point out concepts, tools, and techniques for doing so.

The Magnitude of the Economic Issue in Building

Buildings are expensive. Decisions about investments in buildings usually involve the largest single-item expenditures most people have to deal with during their lifetime, even if they are merely renting an apartment. Expressed in terms of income, an average suburban house costs up to several times an individual wage earner's annual salary. Spread out over time as monthly mortgage payments, it can require anywhere from a quarter to one-half of the monthly family income.

The initial cost of a building, however, appears quite insignificant when compared with the costs incurred to operate and maintain a building over its lifetime. For a 20-year period, these costs can amount to three or four times the initial cost of construction.

Even more impressive is the difference between initial cost and the long-term salary expenditures needed for carrying out the work in a building, for example, of a manufacturing firm. The amount spent on salaries of

1

people working in a building over twenty years can be up to fifty times, or more, of the initial construction cost.

The Challenge to the Architect

One might argue: "Why even look at these issues? The architect cannot do anything about them anyway." Most of the significant factors do seem to be under someone else's control. For example, the architect often is under pressure to speed up the delivery of the building, and thus to cut the time spent on designing and other tasks to the minimum. The client's argument is that "time is money"—every day that the new building cannot be occupied is costing money in interest for loans and lost rental income; just to deliver a workable, feasible, and attractive design solution under such conditions is challenge enough.

Imagine the following situation: It is the crucial meeting between the client and the architect to negotiate the contract for a large building project. The negotiation is almost complete. The client gets ready to sign the contract and, as one of the last points, attempts to get an agreement on the delivery date for the preliminary design.

> *Client:* I expect the preliminary drawings Monday after next—is that OK with you?
>
> *Architect:* Sure. We can get the schematics ready by that date, and you know you will get a professional solution that meets state-of-the-art expectations. But let me make you a little calculation: In the typical project of this kind, about one quarter of our time and fee is spent on what we call "critical design time." That is where the important decisions are made that will determine the project's ultimate success —architecturally, functionally, economically. That is where our firm has earned its reputation for bringing in projects that compare favorably with other projects. But every solution, no matter how good, can still be improved; for example, with respect to economic performance. So let us assume that we could further reduce the initial cost of your project by only 1%, without reduction in quality—that would be $10,000 off your million-dollar project. But we would have to spend a little more time on it—especially in the important schematic design phase. Interested?
>
> *Client:* Mm.
>
> *Architect:* Would you be willing to let us spend, say, $5,000 more on critical design time to achieve this? Over the delivery time of the building, say a year, that would be a 2:1 ratio

of savings in initial cost to the investment of the added design fee.

> *Client:* Not bad, but . . .
>
> *Architect:* At a total A/E fee of, say, 8% (for the sake of simplicity) for the building, which is now going to cost $990,000, that would be roughly $79,000 plus $5,000 = $84,000, which would be about an 8.5% fee. In terms of time, the added effort and cost would be equivalent to about one month in critical design time; two weeks if we put two people on it.
>
> *Client:* Well, I don't know . . .
>
> *Architect:* There is more. If your building requires a total of roughly three or four million dollars in operating, maintenance, energy, and other running costs, over 20 years, and one more month of critical design time could result in a 1% reduction of those costs, this would represent a savings-to-investment ratio of 6:1 or even 8:1 ($30,000 or $40,000 to $5,000). At the same terms as above (aiming at a savings–investment ratio of 2:1), you should be willing to let us spend $15,000 to $20,000 more on design time (five designer months), for a total fee of around 10%. But what if we could actually do it for about the same $5,000 as above, or one more designer month? That would give you a return of $40,000 on an investment of $5,000, which represents an annual return (in savings) of $2,000—that is 40% annually. Which of your stocks gives you that kind of return?
>
> *Client:* Now you are talking. How . . .
>
> *Architect:* Wait; it could get even better. What if we looked at your actual operation and managed to develop a layout that would save you 1% on personnel cost over 20 years? How much do you spend per year on salaries? Fifty people in that size building, at an average annual salary of, say, around $20,000 a year, that is $20,000,000 over 20 years. One percent of that is a savings of $200,000, or $10,000 a year. So, still on the same deal as for the initial cost, you should be willing to spend up to $100,000 more on design costs (that is almost twice the normal fee), and still make $100,000 in savings?
>
> *Client:* Wait a minute . . .
>
> *Architect:* Ok, so what if we could do that for the same $5,000? You would be looking at a return of $200,000 on $5,000 investment over 20 years, or $10,000 each year. That is an annual return rate of 200%.
>
> *Client:* That all sounds too good to be true. Can you show me what kind of things you would

do, or where you have done that in a past project? Can you guarantee the result? Why can't you come up with the better solution right away? Didn't you just argue that "good design is just as cheap as bad design" when you tried to snatch the commission away from your competitors? Besides, you forgot to consider that each month of delay is going to cost me some $7,000 in interest alone . . .

Architect: Well, uh, ah . . .

Client: Have the schematics ready Monday after next. OK?

Of course, this is a hypothetical scenario. But is it implausible or impossible? Most architects would believe that they could achieve those modest reductions in initial cost, annual operation and maintenance costs, even personnel costs where applicable, if they could research the data and spend an appropriate amount of time (even less than what was assumed in the example) to fine-tune the design solution. Is it inconceivable that the architect might arrive at a clever layout and arrangement of functions in the building that might eliminate some unnecessary work or streamline operations so that 1% of salaries could be saved? With all the talk about "form follows function" people have heard from architects, one would expect that this is what they do all the time.

Do we, in the architectural profession, have good, convincing answers to these questions? Clients, especially corporate and government clients, increasingly expect architects to be able to answer them. As they, and their advisors, become more sophisticated in analyzing their financial situation, they expect the architect to follow suit. They know that information as well as analytical methods and tools—computers, programmable calculators, spreadsheet programs, data services —are increasingly available even to the smallest firms: they expect architects to use them. We are running out of excuses for not doing so.

The trouble is that we have rarely looked at our tasks that way, and that is why we would have trouble answering the client's next-to-last question. Backed into the corner, a common reaction on the part of some architects is to let the client negotiate their commission down even further (they do not like to talk about this) —from which the client might conclude that architects really do not know what they are doing in this regard. Is that wrong? Apparently this conclusion is more common than we think; consider the following story.

A number of years ago I was teaching a design class in a Far Eastern city. The project was a large downtown office building, using the site and program of an actual project of this kind. The architect came in to talk to the students, who had spent the first weeks studying efficient access and service core configurations, massing, and similar issues. It was amazing to watch the students' astonishment when the architect casually mentioned that the client had had a team of five financial feasibility advisors work on the project for many months before the architect was involved. None of the advisors was an architect. They gave him a program with the complete service core, number of floors, and outside dimensions worked out in considerable detail, leaving the architect essentially to design a skin around the building.

The message was clear: The client, a major international corporation, did not seem to have sufficient confidence in architects' ability to advise on these crucial feasibility considerations even to make an architect part of this team.

Was this just an isolated incident? The result of some disappointing cost overruns in previous projects? An indication of a general attitude toward architects? A realistic assessment of the (lack of) competence of architects? Many architects would disagree especially with the last judgment—some no doubt quite vehemently, and some even with good justification. But even if the client's attitude were different, how many architects live up to the expectation of reliably carrying out such feasibility analyses, or are competent members of teams such as the one above?

My own architectural education hardly mentioned these issues. As a graduate entering the profession after my first degree, I would have been quite lost had anyone asked me to estimate the cost, let alone the feasibility or financial performance, of a building. I suspect that many graduates of schools of architecture today, even practicing architects, are in a similar situation. It does not have to be this way. In spite of sometimes confusing jargon, and some fierce-looking mathematical equations, it is quite possible to grasp the basics and to make building economics just one more of the many balls the architect has to keep juggling, and even to turn it into an advantage.

Architects like to see themselves as problem-solvers. The economic questions of buildings always are a major part of the client's problem. We cannot afford to ignore a part of the problem the client often considers is the most critical one—even if we ourselves are more concerned with other facets such as aesthetics, user needs, environmental response, or image. Including economic factors in our range of design concerns will not compromise our designs; it will make us better designers. What it takes is, first, a change of attitude.

BUILDING ECONOMICS: SCOPE, RECURRING PROBLEMS

What Is Building Economics?

Once we have accepted the challenge of finding out more about building economics, a reasonable first step is to ask what it is. Can we define the field? Is it meaningful to look for a definition? How does it relate to other fields that also carry the label of "economics" and to economics in general? It turns out that there are quite a few such related fields, and that there is considerable overlap among them. Engineering economics (which has been the subject of many important textbooks), real estate economics, urban economics, energy economics, and environmental economics are a few of the most important examples. Most of these are concerned with what, in general economic theory, would be called microeconomic analysis (the study of how individual actors in the economic realm make economic plans and decisions) as opposed to macroeconomics (the study of laws governing the economy as a whole). Of the fields named above, urban economics and energy/environmental economics are probably most concerned with macroeconomic issues.

These fields share a common base of concepts and techniques that are applied to different areas of practical concern. For this reason, in the past architects with an interest in economics could easily draw their theoretical and technical knowledge from engineering economy or real estate economics textbooks; only recently has building economics emerged as a field of study in its own right.

However, several important directions can be distinguished in building economics. For one, given the importance of the building industry in any national economy, there are important macroeconomic issues that must be studied and would properly fall under the label of building economics: What are the relationships between the construction-related segments of the economy and other fields, and general economic conditions such as those that are influenced by decisions made by the government or the Federal Reserve Board? How can and should areas of vital national importance such as housing be influenced and kept healthy? Such issues have been studied in other areas, such as political economy, sociology, urban policy, transportation, and regional planning. At the other end of the spectrum are the concerns of the practitioners involved with actual and specific building projects, studying construction estimating, construction management, project management, construction financing, and real estate financing. The economics of the various production processes, including transportation processes, would constitute another complex area in building economics. All these fields are further divided into areas of concern by building type; the economics of housing, for example, are quite different from the economics of commercial projects such as office buildings, shopping centers, or industrial buildings.

Building economics issues, even for specific projects, look quite different to the various actors involved. The economic decisions for the owner of a building project are embedded in a range of economic choices going far beyond the building itself, in a way that is of little or no concern to either architect or contractor once the decision has been made to go ahead with a building. The outcomes of decisions by the owner, architect, contractor, and so on, form the context for the economic concerns of the users of buildings, which, by comparison, seem to have received less organized attention, except in a few fields such as housing where the issue has become a main ingredient of larger political-ideological controversies. Marketing studies try to close the gaps between user needs and the products offered by the building industry to respond to those needs, but are applied only where the markets are of sufficient size. In other cases, this concern is assumed to be addressed adequately by the individual rapport between owner, architect, and builder.

Finally, the economics of the process of planning and designing buildings could be seen as a legitimate subfield of building economics that still is waiting for a systematic treatment, even though many people have been concerned with this issue in practice.

This brief survey shows that building economics is far from a well-defined area of study, and that it would probably be counterproductive to insist on a concise definition, which would either exclude some of the concerns mentioned or else turn out so broad and general as to be of little practical value. It also shows that a single book such as this cannot possibly do justice to all the facets of building economics that surfaced even in this quick scan of the terrain. It should be understood that this effort is limited to a particular perspective—that of the architect who tries to get a better grasp of the economic implications of his or her architectural design decisions, and tries to understand how these implications affect others involved in building projects.

The Main Aspects of Recurring Building Economics Problems

There are many different decision and design problems in the realm of building economics, depending on what is known, assumed, or given in a specific situation, and what answers must be found in that situation. Figure 1-1 indicates numerous possible combinations of givens and answers, which can be classified as follows: First is the level at which the decisions are made. The situation that comes to mind most often is that of the individual decision-makers for a specific project trying to arrive at the best decisions for that project—such as the architect trying to design a specific building to best serve the interests of the project client. Building economics issues also include the concerns of specific groups involved in the building process within society or the overall economy, with the aim of safeguarding the interests of those groups and maintaining or im-

GIVEN	FIND
PREVIOUS DECISIONS	DECISION
Go-ahead decision	Go-ahead To next stage
To next stage	Overall project
Overall project	Partial solution /
Solution features	solution feature
	SOLUTION
SOLUTION PROPOSAL(S)	CONCEPT
CONCEPT	PROGRAM Size
PROGRAM	Type
ARCHITECTURAL DESIGN	DESIGN
Schematic design	Schematic design
Design development	Design
	development
Details	Details
DELIVERY PROCESS	DELIVERY PROCESS
FINANCING ARRANGEMENTS	FINANCING ARRANGEMENT
Budget	
Financing conditions	
SCHEDULING / PHASING	SCHEDULING / PHASING
LOCATION / SITE	LOCATION / SITE
CONTEXT CONDITIONS	SUITABLE CONTEXT
A) Given (current)	
B) Future assuming certainty	
C) Future assuming risk	
D) Future assuming uncertainty	
EXPECTED / REQUIRED PERFORMANCE	PERFORMANCE
SOURCE / KIND	
Owner objectives	Acceptability
Regulatory provisions/standards	Feasibility
Market expectations	Satisfactory
Comparative (among solutions)	Best comparative
	Best possible (optimal)
PERFORMANCE MEASURE	
Cost	IMPROVEMENT STRATEGY
Benefit/value	
Period	
Cost/value or cost/benefit	
relationship	

LEVEL OF REQUIRED DECISIONS:
    INDIVIDUAL PROJECT (OWNER, ARCHITECT ETC.)
    GROUP OF PARTICIPANTS (INTEREST GROUP)
    GOVERNMENT OR SOCIETAL LEVEL

FIGURE 1-1.   Constituent aspects of building economics problems.

proving their economic position relative to other parties. This can take two main directions: seeking to maintain or improve the context conditions within which the group operates; or trying to help members of the group become more competitive and effective by developing better methods and techniques for doing its work, providing better information and the analysis tools needed to turn that information into better decisions. Finally, there is the level of governmental or societal policymaking, which is concerned with the overall effectiveness of the entire system of building planning, design, production, operation, and maintenance, and the proper balance among all its parties and components.

Other distinctions concern the nature of the previous decisions, which could be "go-ahead" decisions for proceeding to the next stage of the delivery process, in the implementation of the project as a whole, or could relate to specific features of the eventual solution.

If the givens pertain to solution proposals (one or several), these might be about various progressively detailed stages of planning of the building solution from concept, program, schematics, and so on, to construction details and specifications. But the solution also could entail particular forms of project delivery, contractual or financing arrangements (most prominent here the allotted budget), scheduling or phasing decisions, or the location and choice of site.

Givens always refer to context conditions of all kinds—factors that influence how well a solution to a problem will work but which are not under the problem-solver's or designer's control. Useful distinctions here have to do with assumptions about whether the context is known with certainty (which is really true only if the conditions described are past or present) or involve forecasts. The future usually holds several or many possible context situations, so that forecasts about which outcome might occur are fraught with risk or uncertainty. With respect to the outcome, the set of givens could include performance expectations based on owner goals and objectives, on standards prescribed by laws or regulations, on market expectations, or simply on a comparison among competing solution alternatives. All such expectations are made with tacit or explicit reference to some form of performance measure, which might involve cost (initial or future) only, value or benefit only, the time frame or period in which certain objectives are reached, or one of several measures based on ratios or relationships between combined costs and benefits.

Switching our attention from the givens to the answer(s) that must be found, distinctions again can be made, based on the nature of the decision to be made; deciding to go ahead to the next stage, with the project as a whole, or with adoption of some feature of the eventual solution; or deciding on an aspect of the solution or solutions under scrutiny (concept, program, design, process, location or site, scheduling and phasing, contractual and financing arrangements, etc.). Also, the most suitable context conditions might have to be found for a set of given decisions. Finally, many different analysis techniques have evolved for the different performance aspirations that could dominate the decision: whether the decision merely needs to reach the acceptability or feasibility of a solution, a satisfactory level of performance, the best of a set of competing alternatives, or the optimal—that is, the best possible —solution. Again, all these distinctions are made with reference to some type of measure of performance: cost, benefit period, or some measure including both costs and benefits.

The total number of possible combinations and thus problem types arising from this brief survey is very large; and not all combinations are necessarily meaningful. The following list includes some typical building economics problems that often are encountered:

1. Given a solution proposal and established budget, is the proposal within the budget?
2. Given several solution proposals, which will have the lowest initial cost?
3. Given several solution proposals, which will have the best performance (measured in terms of both costs and benefits)?
4. Given the program, site, and context conditions, find the best possible (optimal) solution.
5. Given the program and performance expectations (cost and other measures), which of several proposed sites is most preferable?

APPROACH

The variety of problems outlined in the preceding section require a considerable range of conceptual, procedural, and technical tools for their proper resolution. There are many ways in which these tools might be introduced and discussed. To preserve some order and sense of progression in this presentation, the main sequence of discussion begins with the initial cost of individual building projects (Chapters 2, 3, and 4), continues with the future costs of operating and maintaining a building over time (Chapter 5), and then takes up the question of benefits or value derived from a project (Chapter 6). More complex measures of performance, involving the relationships between costs and benefits, periods of time in which to achieve the desired performance, and the way in which risk and uncertainty affect performance expectations, are discussed in Chapter 7. Chapter 8 presents a selection of special analysis techniques, using the measures of performance explained in previous chapters. Feasibility analysis is treated in Chapter 9. Chapter 10 attempts to provide a somewhat larger view of building economics issues, going beyond the scope of the decisions involving a single project.

In discussions of these issues, reference is made to many concepts and techniques that may or may not be familiar to all readers, but that cannot easily be fully explained where they first occur without interrupting the flow of the text. A number of these concepts are discussed in some depth in the appendix, in divisions that are organized like the chapters in the main body of the book but do not follow any particular sequence. The appendix should be used as a reference section. An alphabetically organized glossary offers brief explanations and definitions for individual terms and concepts, and a list of references suggests sources for further study.

STUDY QUESTIONS

1. Explain the concept of critical design time, and try to identify periods of critical design time within your own work on current design projects.
2. In Figure 1-1, identify the significant aspects of building economics problems for the sample problems described separately following the figure.
3. Identify additional meaningful combinations of building economics problems that you may have encountered, draw their profile in the figure, and describe them briefly in a coherent narrative.
4. Can you think of other bases for classifying building economics problems? Discuss your answer.
5. In the comparison of Figure 1-2 change the percentage of improvement for each scenario to 0.5% and

	BASE SOLUTION	ALTERNATIVE 1	ALTERNATIVE 2	ALTERNATIVE 3	ALTERNATIVE 4
Personnel needed	200	200	200	200	**198**
Space per person	150	150	150	150	150
Net Leasable area	30000	30000	30000	30000	29700
Net to Gross Ratio	0.8	**0.81**	0.8	0.8	0.8
Total Floor Area	37500	37037	37500	37500	37125
Construction Price	50	50	50	50	50
Initial Building Cost	$1,875,000	$1,851,852	$1,875,000	$1,875,000	$1,856,250
DESIGN COST 2.50%	$46,875	$46,296	$46,875	$46,875	$46,406
Financing LVR	0.8	0.8	0.8	0.8	0.8
LOAN	$1,500,000	$1,481,481	$1,500,000	$1,500,000	$1,485,000
Interest rate	10.00%	10.00%	10.00%	**9.50%**	10.00%
Mortgage term	20	20	20	20	20
Debt service	$176,189	$174,014	$176,189	$170,215	$174,428
Total loan paymnt	$3,523,789	$3,480,285	$3,523,789	$3,404,301	$3,488,551
Financing cost	$2,023,789	$1,980,285	$2,023,789	$1,904,301	$1,988,551
Oper.,maint, R&R rate	$8	$8	**$7.50**	$8	$8
Annual Operat., maint., r&r.	$300,000	$296,296	$281,250	$300,000	$297,000
Total OMR	$6,000,000	$5,925,926	$5,625,000	$6,000,000	$5,940,000
Total Initial, Financing and OMR	$9,898,789	$9,758,063	$9,523,789	$9,779,301	$9,784,801
Savings		$140,726	$375,000	$119,488	$113,988
		1.42%	3.84%	1.25%	1.17%
		Reduction in total 20-year initial and operating cost by improving NGR will justify up to 303.97% increase in critical design cost	Reduction in total 20-year initial and operating cost by red. operating cost will justify up to 800.00% increase in critical design cost	Reduction in total 20-year initial and operating cost by reducing MINT will justify up to 254.91% increase in critical design cost	Reduction in total 20-year initial and operating cost by reducing program requ. will justify up to 245.63% increase in critical design cost

FIGURE 1-2.   Possible increase in critical design time justified by savings.

calculate the resulting cost savings as well as the amount of additional critical design time that could be devoted to the work before half of the savings are used up. Then try 2%. Compare. (See Appendix 4 for an explanation of spreadsheets as calculation and analysis tools.)

2
The Initial Cost of Building Projects

INTRODUCTION

The Initial Cost or First Cost of a building project traditionally has been, and still is, one of the architect's main economic concerns. Most often this takes the form of trying to keep the initial cost within the budget, which is the amount of money the owner has allocated for the project. The budget itself generally is the result of some earlier estimate. In practical terms, the task presents itself in any situation as that of deriving, from the givens at that stage in the building delivery process (see Appendix 2 for a detailed discussion of the building delivery process and its phases), some useful guideline for the decisions required at the next stage. From these decisions, an estimate of the initial cost must be prepared and compared with the budget. If the estimate differs from the budget, either the solution or the budget must be adapted; the problem lies in deciding what changes to make. For example, consider the situation at the programming stage where a preliminary budget may have been established in a prior feasibility analysis. The feasibility analysis may have been done with some rough assumption of how big such a project would be, say, in terms of square feet of total floor area. Together with the tentative budget, this is the given at this stage. The programming work now arrives at a different result for the total floor area and, correspondingly, at a different cost estimate. If this estimate is higher than the previous budget, the programmer must recommend to the client which elements in the program to change to bring the cost back to the given budget—or must get the client to accept a revised higher budget. At the schematic design stage, the program, with its assumption about the total floor area and its program-stage budget, becomes the given. Upon producing a design solution, the designer makes a cost estimate which he or she can compare to this new budget. If the estimate is higher than the budget, will the client agree to the higher cost as the revised budget? Or should the solution be changed? If so, what changes should be considered?

These questions will be discussed in the following chapters. This chapter provides an overview of all the various components that make up the total initial cost of a project. (The most significant of those components from the architect's point of view—the cost of the building—will be examined in some detail in Chapter 3.) A number of different methods for estimating building cost are explained, as well as where and how to find necessary data, the adjustments that must be made for each of these methods, and how the various methods apply to different stages of the project delivery process.

THE ELEMENTS OF INITIAL PROJECT COST: OVERVIEW

The Initial Project Cost is the total amount that must be paid for the planning and construction of a building project up to the point of completion or occupancy. It is useful to think of this amount as what an owner would have to pay at this point if the project were to be paid for in cash, without having to borrow any of it. Adding the cost of borrowing construction funds to that sum (the interest on the amounts borrowed, and the loan fees or "discount points"), we obtain the Total Initial Project Cost.

The first step in making an estimate is to establish what to estimate. What are the various costs items or elements for a building project that together make up the initial cost? In the literature as well as in practice one will find quite different ways of distinguishing the initial cost components: different ways of "slicing the cake." The following list of costs is one such way of distinguishing the most important factors, and Figure 2-1 presents a diagrammatic overview of the initial cost components.

Throughout these discussions, it will be useful to maintain a consistent distinction between *unit prices*, that is, the price of one unit of something (e.g., the price of one square foot of floor material), which often is called the unit cost, and *costs*, which are the result of multiplying the unit prices by the number of units (e.g., the cost of the entire floor of material is the unit price, $/sf, times the number of square feet of floor area). This distinction later is carried into the symbols for the variables, where the suffix CST will stand for cost as defined here, and PRC will stand for unit price. Other than that, and outside this book, these symbols have no special significance.

The cost elements may be defined as follows:

Project cost. The initial cost of the project, including all costs incurred up to the point of completion of the building,

site, landscaping, etc., to have it ready for first occupancy.

Site cost. The cost of acquisition of the site, also called "land cost."

Development cost. The cost of developing the site, that is, clearing the land, providing roads, utilities, planning permits, impact fees, and so on. It may or may not include the fees and profits of the developer, if there is a developer other than the owner involved.

Building cost. The cost of construction of the building itself. However, there may be other construction costs, for example, parking and driveways, runoff control structures, and so forth, which should be listed separately.

Site work cost. The cost of all site preparation and improvements. It includes, for example: landscaping cost; parking cost—the cost of providing surface parking and driveways (underground and multi-story parking are counted as part of the building, and their cost is included in building cost); runoff control cost—the cost of providing measures for stormwater runoff control, such as retention ponds, swales, drainage ditches, and so on; clearing cost—the cost of clearing the site of trees and shrubs as needed. Other items that must be considered as separate cost positions as applicable are extensive leveling/cut and fill work, draining swamps, fences, signage, outdoor furniture, lighting, fountains, pools, and so forth.

Construction cost. Building construction cost and other construction costs on the site (site work) often are lumped together in one common construction cost position or construction budget.

Fees. Fees for various professional services, including: architects' and engineers' fees; legal fees; accounting fees; fees for special consultants' services, such as financial feasibility analysis, marketing, facility programming, and project management, as well as special surveys, soil tests, and, for larger projects, environmental or regional impact studies.

Cost of permits. The cost of permits of various kinds, including the building permit, tree removal permit, development permits, permits for sewage disposal, septic tanks, signage, and so on, as required by applicable regulation.

Carrying charges. The various costs associated with owning, maintaining, and keeping the site in order before and during construction, including such items as: real estate taxes (property taxes); site maintenance; site security cost, which may include fencing, security personnel, and temporary lighting. Management and accounting costs, insurance, temporary utility hookup charges, and the like all would fall in this category, unless specifically included in the construction contracts for site preparation and general conditions.

Interim financing costs (construction financing). Costs that normally consist of interest charges on the construction loan, "discount points" or loan fees, and possibly other fees associated with the process of arranging for financing. At completion of the building, the construction loan together with any remaining interest will be converted into the long-term mortgage or "permanent" financing.

Permanent financing costs. The costs associated with obtain-

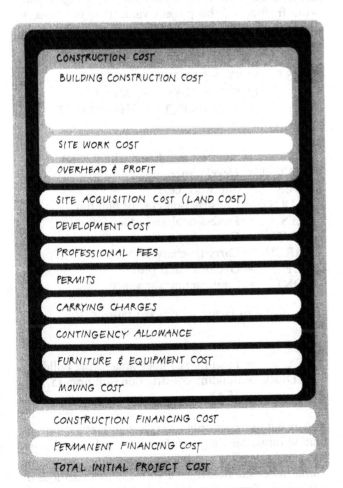

FIGURE 2-1.   Components of initial cost of building projects.

ing permanent financing (usually in the form of a mortgage) for a project. There will be appraisal costs, closing costs, and again "discount points" or loan fees charged by the lender up front. The interest on permanent financing will not be counted in the initial cost.

Overhead and profit. Charges which depend very much on the specific ownership situation for each project, and the relationship arrangements among the owner, developer (if any), general contractor, and subcontractors. Unless the owner acts as the developer, there will be overhead and profit charges for any intermediate firm providing the services of the developer. Even if there is no intermediate developer, for larger projects this will take the form of costs of personnel assigned within the owner's organization to the tasks of managing the project. These costs often are overlooked in early estimates, and if inadequate planning and budget provisions are made for such management, project management difficulties are likely to arise. If separate firms are employed for this purpose, the associated costs might be labeled "management fees" and assigned to the "fees" category.

Equipment cost. The cost of necessary fixed and movable equipment and furniture. A further distinction usually is made between fixed equipment and movable furnishings.

Contingency funds. A certain amount of money that must be set aside for unforeseen costs that may arise during the planning and construction process. The likelihood of such costs arising varies from project to project. If it is a routine one, with standard conditions and construction methods, it can be kept low. As more unknown, experimental, or innovative features are involved, more contingency funds should be set aside. Note that contingency funds should be provided at several levels: first at the building construction level (within the building construction cost and site work cost) and then at the overall project level within the project cost, for unforeseen costs that may arise in areas other than construction.

Moving costs. The cost associated with moving the client's belongings and operations into the new building. This may include not only the costs of the actual moving process but also the costs incurred because of loss of income during the transition.

Total Initial Project Cost. The sum of all initial costs for the project, including interim or construction financing and up-front permanent financing charges.

EQUATIONS FOR TOTAL INITIAL PROJECT COST

The following set of simple equations for the initial costs of a project explains the individual elements in more detail.

$$TIPRJCST = PRJCST + CFINCST + PFINCST$$
(2-1)

where:

TIPRJCST = Total Initial Project Cost

PRJCST   = Project Cost
CFINCST = Construction financing cost
PFINCST = (Up-front) Cost of arranging permanent financing

$$PFINCST = MLOANFEE + CLSCST$$ (2-2)

where:

CLSCST = Closing costs—the various fees and charges associated with arranging for the permanent financing

and:

$$MLOANFEE = MLOAN*DISCPTS$$ (2-3)

Here MLOANFEE or "discount points" is the amount charged by the lender for the mortgage loan (a percentage of the loan).

$$MLOAN = PRJVAL*MLVR$$ (2-4)

PRJVAL is the assessed project value; for new construction it often is assumed equal to the project cost. MLVR is the loan-to-value ratio for the mortgage loan —the fraction of the project value that the bank is willing to lend, with the remainder to be put up by the owner as equity.

$$PRJCST = SCST + DEVCST + SWCST + \\ BLDCST + FEES + PRMCST + \\ CARCH + OHP + CONT.$$ (2-5)

where:
SCST   = Site cost (land acquisition cost)
DEVCST = Development costs
SWCST   = Site work cost
BLDCST = Construction cost of building
FEES   = Professional and other fees
PRMCST = Costs of permits
CARCH   = Carrying charges
OHP   = Overhead and profit
CONT   = Contingency allowance

$$BLDCST = TFA*BLDPRC$$ (2-6)

where:
TFA   = Total (gross) floor area of the building
BLDPRC = Building construction unit price ($/sf of TFA)

This equation represents a very rough method of estimating initial building cost (area method, see Chapter 3). A more detailed breakdown of building cost will be found under the headings of various other methods of estimating first cost; for example, the enclosure method, systems method, and quantity-takeoff-based methods.

$$CFINCST = CLINTAMNT + CLOANFEE \quad (2\text{-}7)$$

where:

CLOANFEE = Loan fee or "discount points" for the construction loan

CLINTAMNT = Amount of interest charged for the outstanding balance of the construction loan

$$CLINTAMNT = AVOB*CLINT*CPER \quad (2\text{-}8)$$

where:

AVOB = Average Outstanding Balance of construction loan

CLINT = Construction loan interest rate

CPER = Construction period

Note: CLINT and CPER must be expressed in terms of the same time units. If CPER is expressed in months, CLINT (usually quoted as an annual rate) must be converted to a monthly rate also.

$$AVOB = CLOAN*AVBALPC \quad (2\text{-}9)$$

where:

CLOAN = Construction loan

AVBALPC = Percentage of the loan assumed to be outstanding, on average, during construction

$$CLOAN = CLVR*PRJCST \quad (2\text{-}10)$$

where CLVR is the loan-to-value ratio for the construction loan, the percentage of the project cost the bank is willing to lend.

Of course, the remaining components of equation (2-5) also must be calculated in detail, according to the applicable conditions of each project. The main concern for the architect is likely to be the building cost, discussed in Chapter 3.

Hard Costs versus Soft Costs

In practice, a distinction is often made between the "hard" costs of a building project as opposed to the "soft" costs. This refers to the distinction made by the tax laws applicable at a given time between costs that must be depreciated over the life of the building (hard costs) and those that must be counted as expenses to the project immediately (soft cost). Because the tax laws in this respect are subject to frequent changes, these costs must be gone over with a tax accountant. The significance of this distinction lies in the fact that it

sometimes determines what can be financed (hard costs) as opposed to what must be paid by the owner.

APPLICATION

The variety of the cost items that make up the total initial project cost for a building project may be surprising to someone unfamiliar with their scope. In dollar amounts, too, these items "add up," as a simplified example of an actual project shows (Figure 2-2). By the time all is done, the total cost is nearly twice that of the building alone; or, in terms of cost per square foot of total floor area, if the construction cost for the building itself is assumed as $55, the cost per square foot including all the other charges is going to be about $100, or more. Because the contribution of these other costs to the total is so significant, the architect must pay close attention to them as well—especially those that directly or indirectly are linked to the architect's decisions, as many are. The connection is not always so easily seen and understood as that of the architect's fees being calculated as a percentage of the construction cost—a connection most clients readily understand, and which for some is a reason, justified or not, for suspecting that the architect is less serious than he or she claims to be about trying to keep construction cost down.

In an effort to understand these connections, we might examine the percentage distribution of the costs in the example. The example should be seen as just that—an example. Some of its cost positions are based on the circumstances of the particular case, such as the land cost, and cannot be used to draw general conclusions, whereas others are arrived at from some analysis of real expected costs, using cost data manuals and other information, such as the construction price or the financing interest rates. They are, however, subject to change over time and by location. Still other items are no more than rough rules-of-thumb, such as the percentages used to determine professional fees, permits, and so on. In reality, they vary with the project type and location, or are subject to a range of different contractual arrangements and even negotiation.

Which items are directly influenced by the architect's decisions? Clearly, the building construction cost will be (at least to the extent that architectural decisions determine overall square footage, building complexity, and so on); and these influences will be more closely scrutinized in Chapter 3. A second group of decisions has to do with trade-offs between land use and building form, sometimes even relating to how much land will be needed. Whether parking is placed on ground or under or inside the building can make a huge difference, for example. Third, all those items that are calculated/estimated as a percentage of building construction cost are clearly but less directly related to the

EXAMPLE FOR TOTAL INITIAL PROJECT COST

I. Top-down model of simplified assumptions for initial project cost

1 Total Inital Project Cost	TIPRJCST	=	PRJCST + CFINCST + PFINCST
2 Permanent Financing Cost	PFINCST	=	MLOANFEE + CLSCST
3 Mortgage loan fee	MLOANFEE	=	MLOAN * DISCPTS
4 Mortgage loan	MLOAN	=	PRJVAL * MLVR
5 Closing cost	CLSCST	=	MLOAN*CLSPC
6 Construction financing cost	CFINCST	=	(CLOAN*AVOBPC*CLINT*CP)+(CLOAN*DISCPTS)
7 Construction loan	CLOAN	=	PRJCST*CLVR
8 Project Cost	PRJCST	=	SCST+DEVCST+CCST+FEES+PERMCST +CARCH+CONT+FURNCST+MOVCST
9 Site acquisition cost (land cost)	SCST	=	SITAR * SITPRC
10 Development cost	DEVCST	=	SITAR*DEVPRC
11 Construction cost	CCST	=	BLDCST+SWCST+OHP
12 Building construction cost	BLDCST	=	TFA*BLDPRC
13 Site work cost	SWCST=	=	PARCST+LSCST
14 Parking and driveway areas cost	PARCST	=	PARKAR*PARPRC
15 Parking and driveway area	PARKAR	=	NOPARKSP* PARSPQ
16 Number of parking spaces	NOPARKSP	=	TFA/PARKREQ
17 Landscaping cost	LSCST	=	LSAR*LSPRC
18 Landscaping area	LSAR	=	SITAR-PARKAR-BFP
19 Building footprint	BFP	=	TFA/NOF
20 Overhead and profit	OHP	=	OHPC*(BLDCST+SWCST)
21 Professional fees	FEES	=	FEEPC*CCST
22 Permits and impact fees cost	PRMCST	=	PERMPC*CCST
23 Carrying charges	CARCH	=	RETX+INSCST+SITSECST+TMPUTCST
24 Real estate taxes (dur. constr.)	RETX	=	SCST*MILLS*CP
25 Insurance during constr.	INSCST	=	INSPC*CCST*CP
26 Site security cost	SITSECST	=	SITAR*SITSECPRC*CP
27 Temporary utilities dur. constr.	TMPUTCST	=	TMPUPRC*CCST*CP
28 Contingency allowance	CONT	=	CONTPC*CCST
29 Furniture and equipment cost	FURNCST	=	FURNPRC*TFA
30 Moving costs	MOVCST	=	MOVPRC*TFA

FIGURE 2-2A.   Spreadsheet example for total initial cost—overall model.

architect's decisions. There are other significant but less obvious influences. One such factor has to do with time—specifically, the total time needed for project delivery and the construction period. If the architect can select construction materials and methods, contractual arrangements, and so forth, that shorten these periods, these choices will reduce the extent of price increases due to inflation between the time of the estimate and project completion; they also will reduce the amount of interest the client must pay for the construction loan.

A good way to study these influences to get a feel for their relative significance is to develop a spreadsheet for the problem such as that shown in Figure 2-2, to begin to vary the different input variables to see how the overall cost changes, and to watch the relative distribution percentages. Figure 2-3 uses the example of Figure

2-2 with several added columns in which selected input variables have been changed.

STUDY QUESTIONS

1. Develop a spreadsheet such as the one shown in Figure 2-2. (See Appendix 3 for a detailed discussion of spreadsheets and their use.) Guess which five or ten input variables are influencing the total cost most significantly, and try to estimate roughly how much the change would be if you varied these variables by, say, 1%, 5%, or 10%. One by one, vary these variables up and down successively by 1%, 5%, and 10%, and note the resulting total cost figures. Compare the results. Which variable had the most profound influence? How does this compare with what you expected? Now check some of the vari-

II. Input variables:

Variable name	(Assumptions) Symbol	In order of appearance in top-down model Amount	Unit
Discount points (percentage)	DISPTS	3.00%	of MLOAN
Mortgage loan to value ratio	MLVR	80.00%	of MLOAN
Closing cost percentage	CLSPC	0.50%	of MLOAN
Project Value (assumed = PRJCST)	PRJVAL	$10,253,573	= PRJCST
Aver. outst. balance percentage	AVOBPC	50.00%	of CLOAN
Construction loan interest rate	CLINT	12.00%	of Average outst. bal ?yr
Construction loan to value ratio	CLVR	80.00%	of PRJCST
Site area	SITAR	300000	sf
Site price	SITPRC	2	$/sf of SITAR
Development price	DEVPRC	0.15	$/SF OF SITAR
Total floor area (program)	TFA	100000	sf
Building construction price	BLDPRC	50	$/sf of TFA
Parking&driveway constr. price	PARPRC	$4.50	$/sf of PARKAR
Parkg area requiremnt (incl. drives)	PARSPQ	300	sf / Parking space
Parking requirement (no. of spaces)	PARKREQ	200	sf TFA/Parking space
Landscaping price	LSPRC	$5.00	$/sf of LSAR
Number of floors	NOF	3	#
Overhead &Profit percentage	OHPC	15.00%	of CCST (assuming OHP not in BLDPRC)
Fees percentage	FEEPC	10.00%	of CCST
Permits percentage	PERMPC	1.00%	of CCST
Real estate taxes ("mills")	MILLS	3.00%	of SCST/YR
Construction period	CP	1.5	years
Insurance percentage	INSPC	0.50%	of CCST
Site security price	SITSECPRC	$0.50	$/sf of SITAR/yr
Price of temp. util. dur. constr.(%)	TMPUTPRC	0.50%	% of CCST/yr
Contingency allowance percentage (:	CONTPC	5.00%	of CCST
Furniture price allowance	FURNPRC	$8.00	$/sf of TFA
Moving cost allowance	MOVPRC	$1.00	$/sf of TFA

III. Calculations: ("bottom-up" model)

			Amount		Percent of total
9	Site acquisition cost (land cost)	SCST	=	$600,000	SITAR * SITPRC	5.21%
10	Development cost	DEVCST	=	$45,000	SITAR*DEVPRC	0.39%
12	Building construction cost	BLDCST	=	$5,000,000	TFA*BLDPRC	43.38%
16	Number of parking spaces	NOPARKSP	=	500	TFA/PARKREQ	0.00%
15	Parking and driveway area	PARKAR	=	150000	NOPARKSP* PARSPQ	1.30%
14	Parking and driveway areas cost	PARCST	=	675000	PARKAR*PARPRC	5.86%
19	Building footprint	BFP	=	33333	TFA/NOF	0.29%
18	Landscaping area	LSAR	=	116667	SITAR-PARKAR-BFP	1.01%
17	Landscaping cost	LSCST	=	$583,333	LSAR*LSPRC	5.06%
13	Site work cost	SWCST=	=	$1,258,333	PARCST+LSCST	10.92%
20	Overhead and profit	OHP	=	$938,750	OHPC*(BLDCST+SWCST)	8.15%
11	Construction cost	CCST	=	$7,197,083	BLDCST+SWCST+OHP	62.45%
29	Furniture and equipment cost	FURNCST	=	$800,000	FURNPRC*TFA	6.94%
30	Moving costs	MOVCST	=	$100,000	MOVPRC*TFA	0.87%
21	Professional fees	FEES	=	$719,708	FEEPC*CCST	6.24%
22	Permits and impact fees cost	PRMCST	=	$71,971	PERMPC*CCST	0.62%
26	Site security cost	SITSECST	=	$225,000	SITAR*SITSECPRC*CP	1.95%
27	Temporary utilities dur. constr.	TMPUTCST	=	$53,978	TMPUTPRC*CCST*CP	0.47%
25	Insurance during constr.	INSCST	=	$53,978	INSPC*CCST*CP	0.47%
24	Real estate taxes (dur. constr.)	RETX	=	$27,000	SCST*MILLS*CP	0.23%
23	Carrying charges	CARCH	=	$359,956	RETX+INSCST+SITSECST+TMPUTCST	3.12%
28	Contingency allowance	CONT	=	$359,854	CONTPC*CCST	3.12%
8	Project Cost	PRJCST	=	$10,253,573	SCST+DEVCST+CCST+FEES+PERMCST +CARCH+CONT+FURNCST+MOVCST	88.97%
7	Construction loan	CLOAN	=	$8,202,858	PRJCST*CLVR	71.17%
6	Construction financing cost	CFINCST	=	$984,343	(CLOAN*AVOBPC*CLINT*CP)+(CLOAN*	8.54%
4	Mortgage loan	MLOAN	=	$8,202,858	PRJVAL * MLVR	71.17%
3	Mortgage loan fee	MLOANFEE	=	$246,086	MLOAN * DISPTS	2.14%
5	Closing cost	CLSCST	=	$41,014	MLOAN*CLSPC	0.36%
2	Permanent Financing Cost	PFINCST	=	$287,100	MLOANFEE + CLSCST	2.49%
1	Total Inital Project Cost	TIPRJCST	=	$11,525,016	PRJCST + CFINCST + PFINCST	100.00%

FIGURE 2-2B.   Initial cost spreadsheet—input variables, calculations.

II. Input variables:

Variable name	Symbol	Amount	Unit			
		(Assumptions)	In order of appearance in top-down model			
Discount points (percentage)	DISPTS	3.00% of MLOAN		4.00%		3.00%
Mortgage loan to value ratio	MLVR	80.00% of MLOAN		80.00%		80.00%
Closing cost percentage	CLSPC	0.50% of MLOAN		0.50%		0.50%
Project Value (assumed = PRJCST)	PRJVAL	$10,253,573 = PRJCST		$10,253,573		$10,253,573
Average outst. balance perc.	AVOBPC	50.00% of CLOAN		50.00%		50.00%
Construction loan interest rate	CLINT	12.00% of Average outst. bal /yr		12.00%		12.00%
Construction loan to value ratio	CLVR	80.00% of PRJCST		80.00%		80.00%
Site area	SITAR	300000 sf		300000		300000
Site price	SITPRC	2 $/sf of SITAR		2		2
Development price	DEVPRC	0.15 $/SF OF SITAR		0.15		0.15
Total floor area (program)	TFA	100000 sf		100000		100000
Building construction price	BLDPRC	50 $/sf of TFA		50		55
Parking&driveway constr. price	PARPRC	$4.50 $/sf of PARKAR		$4.50		$4.50
Parkg area requ. (incl. drives)	PARSPQ	300 sf / Parking space		300		300
Parking requirement (no. spaces)	PARKREQ	200 sf TFA/Parking space		200		200
Landscaping price	LSPRC	$5.00 $/sf of LSAR		$5.00		$5.00
Number of floors	NOF	3 #		3		3
Overhead &Profit percentage	OHPC	15.00% of CCST (assuming OHP not in BLDPRC)		15.00%		15.00%
Fees percentage	FEEPC	10.00% of CCST		10.00%		10.00%
Permits percentage	PERMPC	1.00% of CCST		1.00%		1.00%
Real estate taxes ("mills")	MILLS	3.00% of SCST/YR		3.00%		3.00%
Construction period	CP	1.5 years		1.5		1.5
Insurance percentage	INSPC	0.50% of CCST		0.50%		0.50%
Site security price	SITSECPRC	$0.50 $/sf of SITAR/yr		$0.50		$0.50
Price of temp. util. dur. constr.(%)	TMPUTPRC	0.50% % of CCST/yr		0.50%		0.50%
Contingency allowance percentage	CONTPC	5.00% of CCST		5.00%		5.00%
Furniture price allowance	FURNPRC	$8.00 $/sf of TFA		$8.00		$8.00
Moving cost allowance	MOVPRC	$1.00 $/sf of TFA		$1.00		$1.00

III. Calculations: ("bottom-up" model)

					Percent of total				
9 Site acquisition cost (land cost)	SCST	=	$600,000	SITAR * SITPRC	5.21%	$600,000	5.13%	$600,000	4.89%
10 Development cost	DEVCST	=	$45,000	SITAR*DEVPRC	0.39%	$45,000	0.38%	$45,000	0.37%
12 Building construction cost	BLDCST	=	$5,000,000	TFA*BLDPRC	43.38%	$5,000,000	42.77%	$5,500,000	44.84%
16 Number of parking spaces	NOPARKSP	=	500	TFA/PARKREQ	0.00%	500	0.00%	500	0.00%
15 Parking and driveway area	PARKAR	=	150000	NOPARKSP* PARSPQ	1.30%	150000	1.28%	150000	1.22%
14 Parking and driveway areas cost	PARCST	=	675000	PARKAR*PARPRC	5.86%	675000	5.77%	675000	5.50%
19 Building footprint	BFP	=	33333	TFA/NOF	0.29%	33333	0.29%	33333	0.27%
18 Landscaping area	LSAR	=	116667	SITAR-PARKAR-BFP	1.01%	116667	1.00%	116667	0.95%
17 Landscaping cost	LSCST	=	$583,333	LSAR*LSPRC	5.06%	$583,333	4.99%	$583,333	4.76%
13 Site work cost	SWCST=	=	$1,258,333	PARCST+LSCST	10.92%	$1,258,333	10.77%	$1,258,333	10.26%
20 Overhead and profit	OHP	=	$938,750	OHPC*(BLDCST+SWCST)	8.15%	$938,750	8.03%	$1,013,750	8.27%
11 Construction cost	CCST	=	$7,197,083	BLDCST+SWCST+OHP	62.45%	$7,197,083	61.57%	$7,772,083	63.37%
29 Furniture and equipment cost	FURNCST	=	$800,000	FURNPRC*TFA	6.94%	$800,000	6.84%	$800,000	6.52%
30 Moving costs	MOVCST	=	$100,000	MOVPRC*TFA	0.87%	$100,000	0.86%	$100,000	0.82%
21 Professional fees	FEES	=	$719,708	FEEPC*CCST	6.24%	$719,708	6.16%	$777,208	6.34%
22 Permits and impact fees cost	PRMCST	=	$71,971	PERMPC*CCST	0.62%	$71,971	0.62%	$77,721	0.63%
26 Site security cost	SITSECST	=	$225,000	SITAR*SITSECPRC*CP	1.95%	$225,000	1.92%	$225,000	1.83%
27 Temporary utilities dur. constr.	TMPUTCST	=	$53,978	TMPUTPRC*CCST*CP	0.47%	$53,978	0.46%	$58,291	0.48%
25 Insurance during constr.	INSCST	=	$53,978	INSPC*CCST*CP	0.47%	$53,978	0.46%	$58,291	0.48%
24 Real estate taxes (dur. constr.)	RETX	=	$27,000	SCST*MILLS*CP	0.23%	$27,000	0.23%	$27,000	0.22%
23 Carrying charges	CARCH	=	$359,956	RETX+INSCST+SITSECST+TMPUTCST	3.12%	$359,956	3.08%	$368,581	3.01%
28 Contingency allowance	CONT	=	$359,854	CONTPC*CCST	3.12%	$359,854	3.08%	$388,604	3.17%
8 Project Cost	PRJCST	=	$10,253,573	SCST+DEVCST+CCST+FEES+PERMCST +CARCH+CONT+FURNCST+MOVCST	88.97%	$10,253,573	87.72%	$10,929,198	89.11%
7 Construction loan	CLOAN	=	$8,202,858	PRJCST*CLVR	71.17%	$8,202,858	70.18%	$8,743,358	71.28%
6 Construction financing cost	CFINCST	=	$984,343	(CLOAN*AVOBPC*CLINT*CP) +(CLOAN*DISPTS)	8.54%	$1,066,372	9.12%	$1,049,203	8.55%
4 Mortgage loan	MLOAN	=	$8,202,858	PRJVAL * MLVR	71.17%	$8,202,858	70.18%	$8,202,858	66.88%
3 Mortgage loan fee	MLOANFEE	=	$246,086	MLOAN * DISPTS	2.14%	$328,114	2.81%	$246,086	2.01%
5 Closing cost	CLSCST	=	$41,014	MLOAN*CLSPC	0.36%	$41,014	0.35%	$41,014	0.33%
2 Permanent Financing Cost	PFINCST	=	$287,100	MLOANFEE + CLSCST	2.49%	$369,129	3.16%	$287,100	2.34%
1 Total Initial Project Cost	TIPRJCST	=	$11,525,016	PRJCST + CFINCST + PFINCST	100.00%	$11,689,073	100.00%	$12,265,501	100.00%

FIGURE 2-3. Effect of initial cost changes to example of Figure 2-2.

ables you thought were going to have little or no impact on the total.

2. Using the same spreadsheet, examine the effect of changes in two variables at a time. Given an increase of total cost due to a change in one variable, how great a change in another variable does it take to offset that increase? Try a number of different combinations.

3
Initial Building (Construction) Cost

INTRODUCTION

In the preceding chapter, it was stressed that the cost of construction of a building is only one of many cost factors in the overall initial cost of the project, and that it may not even constitute more than half of the total. Nevertheless, it remains the largest single cost item, and also the one item most directly related to the design decisions of the architect. Thus, without question, the initial building construction cost (building cost) deserves the attention it traditionally is given; nor should we require much convincing that repeating the initial cost estimating of the building at various points during the delivery process is a must. The question then is, how do we do it? Is it just a matter of multiplying building area (square feet) by a dollar-per-square-foot figure? This familiar concept is only the tip of the iceberg. There are many other estimating methods and many different sources of data. What are the differences between these methods? Which ones should we use? How reliable, accurate, and precise are the results? How much effort and what cost are involved in using each of them? These are some of the questions we will study in this chapter.

Concepts discussed in this chapter include: the common underlying concept of all cost estimating methods; cost as opposed to unit price; and types of estimating methods—whole building method, unit-of-use method (order of magnitude estimate), area method, volume method, systems method, trade breakdown method, quantity-takeoff-based methods (unit price Estimates), architects' and owners' estimates versus contractors' estimates. Also considered are: the relationship between the phase of the delivery process and the applica-

bility and usefulness of various methods; modifications and refinements of the methods—according to project size, location, and quality; complexity; the issue of reliability, accuracy, and precision of the estimating methods; and the cost of application.

THE BASIC CONCEPT OF COST ESTIMATING

The variety of methods for estimating the initial costs of building projects may seem confusing at first, and the confusion is compounded by the fact that the names of these approaches vary from author to author. However, there is one simple common underlying idea:

$$\text{COST} = \text{AMOUNT*UNIT PRICE} \qquad (3\text{-}1)$$

or more specifically:

$$\text{Estimated COST} = \text{Estimated AMOUNT} \\ \text{(number of units)} \times \text{estimated} \\ \text{UNIT PRICE} \qquad (3\text{-}1A)$$

Note the use of the term COST for the performance variable to be calculated or estimated. It is the result of the calculation. The term PRICE is used for the unit price; this is the context variable whose value is not under the designer's control but depends on the market or on others' decisions. The AMOUNT is the direct or intermediate design variable, controlled directly or indirectly by the designer.

The equation is conceptually the same as it would be for a final account or bill for the building, except that in an estimate the amounts or the unit price, or both, are not definitively known but only are predicted or estimated. As long as the quantities appearing on the right-hand side of the equation are estimates, the resulting cost figure will be an estimate too.

This is the common basis for all methods of cost estimating. Variations in the methods are due to the choice of different units of measurement for the amounts, and consequently, of the corresponding units of measurement for the unit prices. The methods of estimating can therefore be classified according to the unit of measurement used.

ESTIMATING METHODS FOR BUILDING COST—A SURVEY

Some of the most important types of methods for estimating initial building cost are as follows.

Whole Unit Method (Whole Building Method). This is, strictly speaking, not an estimating method as such because there is only one unit. The unit referred to is the whole building—for example a house (one-family residence) or a school, assuming some standard type and size. The costs quoted in data sources using this method usually are total initial project cost, not mere building construction cost, and it is not always clear which of the two concepts is meant.

Unit-of-Use Method (Order of Magnitude Estimate). Many building types are characterized by repetition of units of use or user stations. For example, the size of hospitals often is expressed in terms of the number of hospital beds. Cost then is estimated by using a unit price per bed. Other examples are schools (student stations), office buildings (office worker stations/desks), and even airports, whose cost can be estimated by the

FIGURE 3-2.   Unit-of-use method.

number of boarding gates and the price per gate. Of course, the unit price in each case includes the cost of the shared common service facilities. This method is used in very early stages of the project, and as a basis for comparison with other projects of the same type and overall scope. This type of estimate also is referred to as an order of magnitude estimate. Again, care must be taken to ascertain whether data quoted refer to building construction costs or total project cost.

Area Method (Square Foot Method). This is one of the most common estimating methods, and is based on the unit of total floor area or gross floor area, for example, as defined by AIA document D 101. Estimating by price per square foot is most common in the United States. Most European and many overseas countries use the unit of square meters—but often do not use the area method as the predominant cost estimating method.

Volume Method (Cubic Foot Estimate). The size of the project here is measured not by floor area but by units of space volume, for example, cubic feet or cubic meters. This method, using cubic meters, is more common in the United Kingdom and Europe than in the United States.

Enclosure Method. An issue of principal interest to the architect is the relationship between cost and building form. Because neither the area method nor the volume method is sensitive to differences in the geometry of buildings, a number of efforts have been made to develop estimating methods that do consider the effect of building geometry and the relationship between the surface of buildings and the enclosed space—hence the name "enclosure method." Some of the approaches that use this name actually are modifications of the

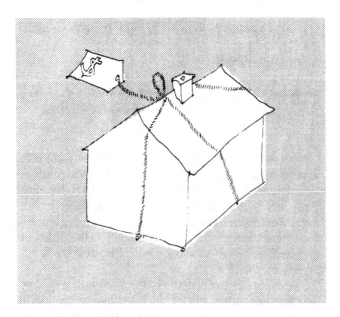

FIGURE 3-1.   Whole unit method.

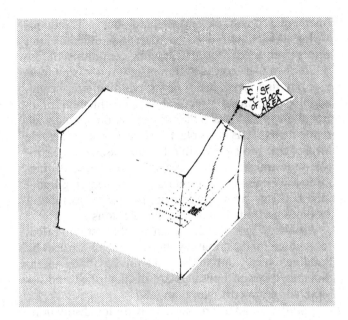

FIGURE 3-3.   Area method.

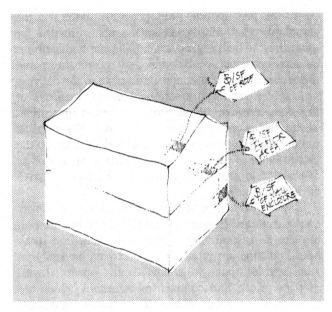

FIGURE 3-5.   Enclosure method.

area method. The method we will use for this purpose considers area units of the actual building surface, roof area, external walls, and so forth, in addition to floor space, and corresponding unit prices.

Systems Method. The base unit here is the same as for the area method, but listed separately for each category of subsystems of the building; for example, structure, mechanical system, foundations, and so on. For each system, unit prices are given as $ per square foot of total floor area, and also as percentages of area method unit prices. The estimate for the foundation may be listed as $5/sf of TFA and as 7% of the total building

cost. Thus, the overall area method unit price is $71.43/sf of TFA.

Trade Breakdown Method. This is conceptually similar to the systems method, but here the basis for the breakdown into categories is not the subsystem but the share of the work done by the different trades participating in the construction of the building.

Quantity-Survey-Based Methods (Unit Price Estimates). A number of variations of estimating methods are based on actual counts (the quantity survey) of various items, materials, components, and so forth, to be used in the building. The unit referred to in the other

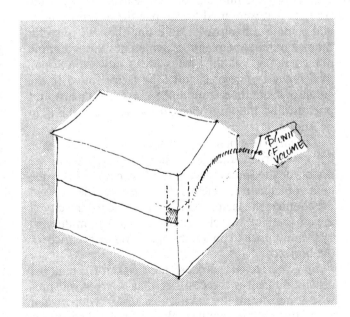

FIGURE 3-4.   Volume method.

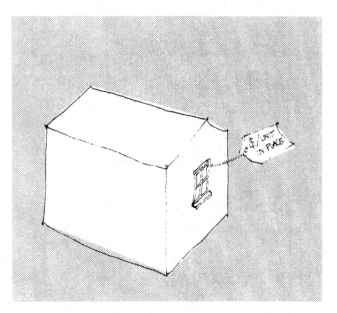

FIGURE 3-6.   Unit-in-place method.

name for this method (unit price estimating) is the actual physical building part, or the measuring unit for quantities of building material. This method requires detailed construction drawings and specifications and therefore cannot be used until quite late in the delivery process. Important distinctions between methods of this type consider the question of whether unit prices are quoted as in-place prices (that is, as finally installed in the building) or as suppliers' shelf prices to which the cost of labor in the installation of each item, as well as overhead and profit, would have to be added to arrive at in-place prices.

Owners' and Architects' Estimates versus Contractors' Estimates. Because of the difference in focus on in-place versus suppliers' prices plus labor and overhead, a distinction often is made between owners' and architects' estimates, which are mainly concerned with in-place prices, and contractors' estimates, which are concerned with the relationship between materials, labor, and overhead costs. Of course, such distinctions could be made for other estimating methods as well, but they usually are applied to quantity-takeoff-based methods.

ADJUSTMENTS AND REFINEMENTS

Each of the estimating methods listed above can be refined in various ways to allow for more specific distinctions and therefore more precise estimates. The re-

FIGURE 3-7. Contractor's estimate.

finements consist of making additional distinctions within the chosen units, or applying different unit prices "adjusted" for the specific conditions of the building under consideration. Common distinctions are: building type; quality level; building size, height, and locality; main structural system; different functional areas within a building; and so on. Using such refinements, it is possible to achieve a considerable level of fine-tuning even in estimates based on the simple and straightforward area method. These distinctions will be especially important in considering trade-offs between alternative materials or construction methods and other major design decisions.

Building Type. The data manuals that list unit prices for area method estimating almost always list prices by building type. Commonly, data for 25 to 40 or 50 different types of buildings are distinguished and assembled in the cost data manuals.

Quality. In addition, most provide for distinctions according to quality levels, for example, by listing low (or ¼), median, and high (or ¾) unit prices. The quartile figures mean that in the sample of projects surveyed to obtain the data, 25% of the projects had unit prices at or below the figure given under the ¼ column, 50% were at or below the price listed under median, and 75% were at or below the prices given for ¾; conversely, this means that 25% of the projects had unit prices higher than the figure given in the ¾ column. This allows an owner to specify whether the project under consideration should be seen as an economy, standard, or luxury one and to apply the appropriate unit prices for the estimate.

Size. Because of the phenomenon of economy of scale, large buildings tend to have lower unit prices than very small projects. To account for this, the data sources provide information on the average size of the projects that were used to compile the unit price data, and some mechanism for adjusting the unit price for the size of building planned. Sometimes this is a graph of a curve (as, e.g., the Means Area Conversion Scale), sometimes a table of factors. The factor found in the graph or table then is multiplied by the standard unit price to find the unit price for the size of building in question.

Locality. To obtain a reasonably accurate estimate, it is necessary to adjust the national or regional average prices for the specific location. Most manuals provide tables of adjustment factors for the major cities around the country; the factor found in the table for the city of the project (or the closest nearby city) is multiplied by the unit price to obtain the appropriate unit price for the project.

Building Height. Because the height of a building usually influences the cost, the data sources will provide some means for taking this into account when estimating. Some manuals already make distinctions for height within the building type (for example, Means distinguishes low-rise, mid-rise, and high-rise apartment buildings and office buildings), and others offer additional adjustment factors to be applied after reading the building type unit price from the tables. In spreadsheets designed to assist the designer in exploring different solutions, it may be advantageous to use (and, if necessary, to construct) an equation in which the height appears as a variable that can be changed as the design solution changes. For example, the effect of height can be expressed as an exponential function as follows:

$$PRC_n = PRC_1*(1 + hf)^n \qquad (3\text{-}2)$$

where PRC_n is the desired price for a building with n floors, hf is an adjustment coefficient (the rate at which the price rises with the number of floors; to be established from a given set of data), and n is the number of floors.

Figure 3-8 shows a spreadsheet where hf is manipulated and adjusted until the curve of height unit price falls within the three benchmark prices given by Means

for low-rise, mid-rise, and high-rise buildings of the same type.

BUILDING COST ESTIMATING BY AREA METHOD

Among the different cost positions making up the overall cost of a project, not only is the element of building cost usually the most significant in terms of magnitude, but it also is the one that commands the most interest on the part of architects, as it is directly related to the building design decisions for which architects are responsible. Of the several approaches available for estimating the initial cost of the building, the area method is the one most commonly used in the early stages of the project.

This section presents a detailed discussion of the area method, together with its typical uses and refinements or adjustments for building type, size, locality, quality, and height. The area method is an approach to estimating the initial cost that describes the building in terms of the total floor area provided. The total floor

BUILDING CONSTRUCTION PRICE AS A FUNCTION OF HEIGHT

FACTOR = $(1+RATE)^n$
BLDPRC(n) = BLDPRC(1)*FACTOR

Number of floors n	Rate hf: 0.0238	BLDPRC(base): 46.2	Rate hf 0.0237	BLDPRC(base) 46.21	
	FACTOR:	BLDPRC(n)	FACTOR:	BLDPRC(n)	MEANS
1	1.0238	$47.30	1.0237	$47.31	
2	1.04816644	$48.43	1.04796169	$48.43	Low-rise
3	1.073112801	$49.58	1.072798382	$49.57	$49.60
4	1.098652886	$50.76	1.098223704	$50.75	
5	1.124800825	$51.97	1.124251605	$51.95	
6	1.151571084	$53.20	1.150896369	$53.18	Mid-rise
7	1.178978476	$54.47	1.178172612	$54.44	$54.60
8	1.207038164	$55.77	1.206095303	$55.73	
9	1.235765672	$57.09	1.234679762	$57.05	
10	1.265176895	$58.45	1.263941672	$58.41	
11	1.295288105	$59.84	1.29389709	$59.79	
12	1.326115962	$61.27	1.324562451	$61.21	
13	1.357677522	$62.72	1.355954581	$62.66	
14	1.389990247	$64.22	1.388090705	$64.14	High-rise
15	1.423072015	$65.75	1.420988454	$65.66	$65.70
16	1.456941129	$67.31	1.454665881	$67.22	
17	1.491616328	$68.91	1.489141462	$68.81	
18	1.527116796	$70.55	1.524434115	$70.44	
19	1.563462176	$72.23	1.560563203	$72.11	
20	1.600672576	$73.95	1.597548551	$73.82	

FIGURE 3-8.   Factors for building cost increase with height.

area is the sum total of all the floor area in the building, measured from the outside of the building exterior walls. [It will be referred to as TFA in the following; see Appendix 1 for a more detailed discussion of the total floor area and the various definitions of net (leasable, rentable, usable) areas, and the relationship between them, e.g., the Net to Gross Ratio, NGR.]

The relationship between Total Floor Area (TFA) and Net areas—usually Net Leasable Area (NLA)—plays an important role in the use of this method. Specifically, the Net to Gross Ratio (NGR) formed by NLA/TFA (or its inverse, the Gross to Net Ratio, GNR: TFA/NLA) is widely used both as an indicator of floor plan efficiency and to establish the probable size of the overall building when only the program in terms of the required NLA is given. For each building type, there are standard guidelines for the average or acceptable NGR. The required NLA of a program is divided by the NGR to obtain the overall TFA. The TFA then is multiplied by the appropriate unit price, BLDPRC (for example, in $ per square foot of TFA), to calculate the estimated building cost:

$$BLDCST = TFA*BLDPRC \qquad (3-3)$$

A few examples illustrate the use of the area method for different situations in the planning of a building, including adjustments and refinements:

1. The program for a hypothetical office building in Miami, Florida calls for 100,000 square feet of Net Leasable Area. How much will the building cost? The only available data manual is from last year; the planning is estimated to take two years. The end of the construction period thus will be up to three years later than the last available unit price information. The building should be of high quality; the site allows for on-ground parking and then will leave an area of about 20,000 sf for the building footprint. The data manual lists $55, $68, and $80/sf of TFA, respectively, for low-, median-, and high-quality office buildings. The median range for this type of office is given as 52,000 sf. First, TFA must be found; the standard NGR for office buildings is given as 0.80 or 80%, and 100,000 divided by 0.8 is 125,000 sf TFA. The height of the building will be at least 125,000/20,000 = 6.25 floors or more; we will assume seven floors. Low-rise, mid-rise, and high-rise prices are given as $61, $80, and $95, respectively, in this category; this would correspond to a height increase factor of about 4% per floor. Applying an exponential factor for seven floors:

$$\$61*(1.04)^7 = \$80.27$$

which would correspond closely enough to the mid-

rise figure given in the manual. The locality adjustment factor for Miami is 0.84; thus the unit price should be adjusted as:

$$\$80.27*0.84 = \$67.43 \text{ per squ. ft. of TFA}$$

The building size is 125,000 sf. However, the data manual lists 52,000 sf as the typical average size of the samples from which the unit prices were developed. Using the Means Area adjustment factor diagram, we calculate first a project size factor of 125,000/52,000 = 2.4, and read from the graph an approximate cost multiplier of 0.93. So, the unit price for the project will be $67.43*0.93 = $62.71/sf. The total cost of the building will be 125,000*62.71 = $7,838,737, or roughly $7,840,000—that is, if the building were to be built during the year for which the manual provided its data. We still have to allow for price increases due to inflation for the three years between the manual's data and the expected completion date. If inflation runs at an annual rate of 5%, we must calculate the compound amount factor of $(1.05)^3 = 1.16$ and multiply it by the above result to get the final expected cost: $9,075,780.

Remember that this kind of early estimate can have an error range of up to ±15%, which means that the final cost could lie anywhere between 7.7 and 10.4 million dollars. It therefore is not meaningful to calculate these figures to a very high degree of precision.

2. The client states that there will be only $6,900,000 available for the initial building cost. How much building and how much NLA can be built for that amount? The above figures should be used with some latitude, because the size adjustment and consequently the height adjustment will be slightly different. First, let us get the budget back to the time of the manual: $6,900,000/1.16 = roughly $6,000,000. Then divide by BLDPRC to get the affordable TFA:

$$\$6,000,000/62.71 = 97,229 \text{ sf TFA}$$

Applying the Net to Gross Ratio: 97,229*0.8 = 77,783 or approximately 77,000–78,000 sf of NLA. This is a first rough estimate of the affordable net area. We can use it to go over the above process again to cross-check our figures: Assuming a TFA of around 100,000 sf divided by a 20,000 sf building footprint yields a building height of five floors. The corresponding height adjustment factor now is 1.22:

$$\$61*(1.04)^5 = \$61*1.22 = \$74.22$$

The size adjustment factor will be 100,000/52,000 = 1.92; from the graph we read a factor of

0.95; thus the national average price is $70.51. Applying the Miami locality factor of 0.84 again, we obtain a square foot price of $59.23; the building cost will be 100,000*59.23 = $5,923,000. This is just below the stipulated budget (before inflation). If we apply the construction price of $59.23 to the $6,000,000 budget, we obtain an affordable TFA of 101,304 sf. Multiplying this by the 0.8 NGR factor, we find that the affordable net area will be 81,043 sf or roughly 81,000 sf.

THE ENCLOSURE METHOD OF ESTIMATING INITIAL BUILDING COST

In this section, we will discuss an approach to estimating initial cost of buildings that is based on the geometry of the design solution. Called the enclosure method, it aims at filling the gap between the area method and more detailed estimating approaches that require information about design decisions that have not been made at the stage where such estimates are needed most.

Enclosure Method: Concept and Aim

One of the shortcomings of the area method is that it is insensitive to differences in architectural form. As long as two schematic design alternatives have the same floor area, the initial cost estimate will produce the same result, even for two solutions as different as the buildings shown in Figure 3-9.

This is a serious problem when one considers that decisions made early in the design process have the most far-reaching consequences on the cost and economic performance of the project. It would be helpful to have an estimating method that would provide quick feedback on the probable economic performance of solutions considered at this stage, so that many alternative solutions could be analyzed before one proceeded to the next stage with a commitment to a particular solution. Moreover, it should be possible to analyze these solutions even before detailed decisions about materials, construction methods, and so on, are finalized. But the standard estimating methods currently require going to a much more detailed level before comparisons of alternative schemes can be carried out. There are some approaches that involve using area method estimates with modification factors based on the perimeter length of the building, but these are crude approximation methods at best.

An approach that could fill this gap is the enclosure method, which is based on a breakdown of the building into its geometric envelope components—roof surface, exterior walls, foundation or contract area of the building with the ground. When these elements are stripped from the building, the basic structure and floor area (including services) remain. The latter can be treated as it would in the area method; however, for the other components units of measurement would be used that are based directly on the envelope surface. The total building cost is the sum total of the costs of the stripped-down floor area and the individual enclosure areas. For example:

$$\text{BLDCSTE} = \text{TFAECST} + \text{RFECST}$$
$$+ \text{EWALLECST} + \text{IWALLECST} + \text{FNDECST}$$
$$+ \text{SCECST} \quad (3\text{-}4)$$

where:

TFAECST	= Cost of the (enclosure) floor area
RFECST	= Cost of the roof enclosure area
EWALLECST	= Cost of the exterior wall enclosure
IWALLECST	= Cost of interior walls
FNDECST	= Cost of the foundation enclosure (building contact with ground)
SCECST	= Cost of service core enclosure component

Each of these costs is estimated by multiplying the respective areas by appropriate surface unit prices. Currently available building cost data are not compiled and prepared in a format that could be used directly for such an approach. They either must be prepared in a roundabout fashion from more detailed data such as in-place item unit prices or compiled from available examples. If they are determined, however, simple mathematical models of building geometry implemented on computer spreadsheets can be prepared and used for quick analysis of the cost implications of a large number of alternatives.

An example of a spreadsheet model of an enclosure method estimate for a simple office building is shown in Figure 3-10.

THE SYSTEMS METHOD OF ESTIMATING INITIAL BUILDING COST

As the design of the building progresses into the design development phase, more information about materials,

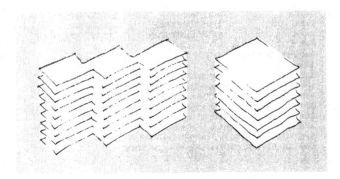

FIGURE 3-9.   Two buildings with the same floor area.

EXAMPLE: SPREADSHEET FOR OFFICE BUILDING SCHEME A (DOUBLE-LOADED CORRIDOR)

Input variables: Design variable values show **bold**

Name	Symbol	Value	Unit of measurement	Source; Constraint	
Net leasable area	NLA	50000 sf		Program	
Module area	MAR	500 sf		Program	
Module length	ML	**15** lf		Design	ML ≥ 10'
Corridor width	CW	**6** lf		Design	CW ≥ 6'
Wall thickness	WT	**0.8** lf		Design	
Building construction price	CPRC	53 $/sf of TFA		Context	
No. modules per floor	NMPF	**10** #		Design	
Floor-to-floor height	FHT	**10** lf		Design	

Equations:

	Variables	Symbol		Value Unit	Formula
8)	Module width	MW	=	33.33 lf	MAR / ML
7)	Number of modules	NOM	=	100 sf	NLA / MAR
6)	Number of floors	NOF	=	10	NOM / NMPF
5)	Building length	BL	=	234.93 lf	2*WT + MW*(NMPF + 4)/2
4)	Building width	BW	=	39.2 lf	2 * (ML + 2 * WT) + CW
3)	Floor Area (one floor)	FA	=	9209.39 sf	BW * BL
2)	Total Floor area	TFA	=	92093.87 sf	FA * NOF
1)	Building Construction cost (Area method)	BLDCST	=	$4,880,975	TFA * CPRC
	(Building construction cost is assumed to include tenant improvement costs)				
9)	Net to Gross Ratio	NGR	=	0.54	NLA/TFA

TOTAL INITIAL PROJECT COST - ENCLOSURE METHOD
Top-down model:

		Symbol		Formula
40)	Building constr. cost -E	BLDCSTE	=	SPCST+RFCST+FNDCST+EWARWCST+EWARQCST+ IWALLACST+IWALLBCST+ELEVCST+ADDLNDCST
41)	Space cost (Fl. area)	SPCST	=	TFA*SPPRC
		SPPRC		Floor area unit price (enclosure)
42)	Roof cost	RFCST	=	RFAR*RFPRC
43)	Foundation cost	FNDCST	=	FNDAR*FNDPRC
	(simplified)	FNDPRC		Foundation unit price (contact with ground)
44)	Exterior wall cost	EWALLCST	=	EWARWCST+EWARQCST
		EWARWCST		Cost of exterior wall - windows
		EWARQCST		Cost of opaque portion of ext. wall
45)	Ext. wall cost-windows	EWARWCST	=	EWARW*EWARWPRC
46)	Ext. wall area - glazed	EWARW	=	EWAR*GOR
		GOR		Glass to opaque wall area ratio
47)	Exterior wall area	EWAR	=	2*BH*(BL+BW)
48)	Building height	BH	=	NOF*FHT
49)	Cost of opaque port. of ext. wll	EWARQCST	=	EWAR*(1-GOR)*EWARQPRC
50)	Cost of int. walls/partitions	IWCST	=	IWCSTA+IWCSTB
51)	Cost of int. wall type A	IWCSTA	=	IWARA*IWPRCA
		IWPRCA		Unit price for wall type A
52)	Interior wall area A	IWARA	=	FHT*(NOMPF+2)*NOF
53)	Cost of interior wall type B	IWCSTB	=	IWARB*IWPRCB
		IWPRCB		Unit price for interior wall type B
54)	Interior wall type B area	IWARB	=	2*BL*FHT*NOF
55)	Cost of elevators	ELEVCST	=	ELEVNO*ELEVPRC+(NOF-2)*ELEVNO*ADDLANDPRC
		ELEVNO		Number of elevators
		ELEVPRC		Unit price per elevator
56)	Cost of additional stops >2	ADDLNDCST	=	ELEVNO*(NOF-2)*ADDLNDPRC
		ADDLNDPRC		Price for each additional landing >2

INITIAL BUILDING COST - ENCLOSURE METHOD - CALCULATIONS

	Variable	Symbol	Value	Unit	Formula	Unit price	Unit	Cost	Symbol
41)	Enclosure model floor area	SPTFA	92,094	sf	= TFA	37.00	$/SF OF TFA	$3,407,473	SPCST
42)	Roof area (assumed flat roof)	RFAR	9,209	sf	= FA	7.00	$/SF OF RFAR	$64,466	RFCST
43)	Foundation/footprint area	FNDAR	9,209	sf	= FA	5.00	$/SF OF FNDAR	$46,047	FNDCST
48)	Building Height	BH	100	lf	NOF*FHT				
47)	Exterior wall area	EWAR	$54,827	sf	BH*2*(BL+BW)*NOF				
	Glass/opaque wall ratio	GOR	0.25		Design				
46)	Window area (glass)	EWARW	13,707	sf	EWAR*GOR	20.00	$/SF OF EWARW	$274,133	EWARWCST
49)	Opaque wall	EWARQ	41120	sf	EWAR-EWAR	16.00	$/SF OF EWARQ	$657,920	EWARQCST
51), 52)	Interior wall area A (betw. mod.)	IWALLA	15,060	sf	4*ML+(NMPF	2.00	$/SF OF IWALLA	$30,120	IWALLACST
53), 54)	Interior wall area B (Corr.)	IWALLB	33,333	sf	NMPF*MW*Fl	2.50	$/SF OF IWALLB	$83,333	IWALLBCST
55)	Elevators	ELEVNO	3	#	Design	42,000.00	$/ELEVATOR	$126,000	ELEVCST
56)	Landings > 2	ADDLND	8	#	NOF - 2	7,000.00	$/LANDING	$168,000	ADDLNDCST
40)	Building Constr.Cost Encl.	BLDCSTE	$		SPCST+RFCST+FNDCST+EWARWCST+EWARQ IWALLACST+IWALLBCST+ELEVCST+ADDLNDCST			$4,857,492	BLDCSTE

FIGURE 3-10. Enclosure method estimate—spreadsheet.

structural systems, and so on will become available and can be drawn upon to keep track of the initial building cost. An approach that conceptually falls between the area method and itemized unit price methods is the systems method, which distinguishes various functional systems or subsystems in the building and develops a separate estimate for each of them. This section discusses this method and its uses. It assumes that the reader is familiar with the area method of cost estimating; it will be useful to have a cost data manual at hand.

A good explanation and strong arguments in favor of the systems method are presented in H. Swinburne's book *Design Cost Analysis for Architects and Engineers*. A building is conceived as consisting of various subsystems distinguished according to their main function. Swinburne advocates distinguishing 15 such systems: foundations, floors on grade, superstructure, roofing, exterior walls, partitions, wall finishes, floor finishes, ceiling finishes, conveying systems, specialties, fixed equipment, HVAC, plumbing, and electrical. This is how the Dodge construction systems cost data manuals are organized.

When the basic principle of cost estimating was introduced above, it was stated that for each method of estimating, we must identify the unit of measurement to which the unit prices will be applied. With the systems method, things become more complex than they were in the methods discussed so far. There are two ways of representing the information of a systems estimate and thus of providing the data needed. On the one hand, design features of a subsystem are analyzed almost as they would be for an itemized unit price estimate. However, the costs and subtotals for the subsystem components are given as costs per square foot of floor area (not as item unit costs), and then summed up to overall system costs expressed as cost per square foot of total floor area. On the other hand, these costs also are expressed as percentages of the overall cost per square foot of building for a particular building type. (The floor-area-based unit prices led Means to subsume the method under "square foot estimating.")

Data for this approach are given for the assumption of an average building with specified assumptions for each system. For example, structural systems data are based on average span and load assumptions. The resulting estimates must be modified by using adjustment factors provided for each system if the planned building design deviates from these assumptions. The method allows an estimate to use fairly standard overall data assumptions for all parts or systems of the building that are not yet known in sufficient detail (but are not expected to deviate much from the norm), and to study particular subsystems with special features or conditions in more detail as needed. The overall estimate then can be adjusted with this more specific information. Figure 3-11 shows an example of a systems method breakdown of a cost estimate.

The systems method is the last estimating method discussed in any detail here, but not because other approaches, especially more detailed versions of square foot estimating and unit price estimating (quantity-survey-based estimating methods), are less important—quite the contrary. But these other methods are less applicable at the early design stages that are the focus of interest here, and they have been described in great detail elsewhere.

OFFICE BUILDING - SYSTEMS METHOD ESTIMATE

Overall building cost (same as Area Method estimate)		BLDCST	$4,880,975	FROM AREA METHOD CALCULATION
Subsystem costs	Percentage of total		Cost	$/sf of TFA
Foundation system cost	6.00%	FNDSYSCST	$292,858	$3.18
Floors on grade cost	1.50%	FLGRDCST	$73,215	$0.80
Superstructure system cost	19.00%	SPSTRCST	$927,385	$10.07
Roofing system cost	2.50%	RFSYSCST	$122,024	$1.33
Exterior wall system cost	16.00%	EWLSYSCST	$780,956	$8.48
Partition system cost	7.50%	PARTSYSCST	$366,073	$3.98
Wall finishes systems cost	5.50%	WFINCST	$268,454	$2.92
Floor finishes systems cost	3.00%	FLFINCST	$146,429	$1.59
Ceiling finishes systems cost	2.50%	CLFINCST	$122,024	$1.33
Conveying systems cost	2.00%	CNVSYSCST	$97,619	$1.06
Specialties systems cost	1.50%	SPECSYSCST	$73,215	$0.80
Fixed equipment systems cost	1.50%	FXEQSYSCST	$73,215	$0.80
HVAC Systems cost	17.00%	HVACSYSCST	$829,766	$9.01
Plumbing systems cost	6.50%	PLMBSYSCST	$317,263	$3.45
Electrical systems cost	8.00%	ELSYSCST	$390,478	$4.24
	100.00%			

FIGURE 3-11. Systems method estimate.

USEFULNESS OF ESTIMATING METHODS AT DIFFERENT PROCESS STAGES

The methods of estimating initial cost listed in the preceding brief survey are based on widely differing assumptions regarding the precision and the information detail needed to actually perform the estimate. These requirements determine the usefulness of each method for a given stage in the delivery process.

The chart shown in Figure 3-12 gives a rough overview of the applicability of the various estimating methods to different project phases.

RELIABILITY; ACCURACY VERSUS PRECISION OF ESTIMATING METHODS

Negative publicity about architects' work with public projects often has to do with cost overruns, that is, with projects ending up costing more than estimated. How reliable are the estimating methods? The question may actually be misstated: the reliability of an estimate depends not just on the method, but even more so on the quality of the data used. In early stages in the delivery process, the available data—with regard to amounts needed, kind of units, and unit prices—are quite approximate and imprecise. Therefore, great expectations about the reliability of the resulting estimates are not appropriate even when the calculation results look quite precise. As the planning progresses, more is known about the amounts and types of building components, and more specific unit prices can be applied. This is why different methods are needed. Estimates can be said to be reliable if they generally turn out to be accurate.

Accuracy often is confused with precision; the terms do mean different things. The statement "It is now about midnight" is accurate if it is actually about midnight when the statement is made. However, it is not very precise. The statement "It is now 11:58 and 4 seconds P.M." is precise, but it may not be accurate—for example, if it is actually 12:07 A.M. The first statement, "about midnight," by comparison, still would be reasonably accurate. The second statement is accurate and precise only if it is actually made at 11:58 and 4 seconds P.M. Accuracy concerns how a statement or datum relates to truth or actual fact; precision merely concerns how finely distinctions are made. Our key concern about accuracy in estimating is about how the estimate relates to the budget set aside for the project, or to the actual cost as given by the selected bid—or by the final accounting after completion. We can measure the accuracy of an estimate by its closeness to that actual cost. Its precision refers to how fine-grained or detailed it is. We usually hope, of course, that making an estimate more precise also will make it more accurate.

Early in the process, an estimate may be considered good if it is within around 15% of actual cost or budget.

PHASE >>> METHOD	OWNER / ARCHITECTS' ESTIMATES >>>>> FEASI- BILITY	PROGRAM PHASE	SCHEMATIC DESIGN	DESIGN DEVELOP- MENT	CONSTR. DOCU- MENT	CONTRACTOR'S ESTIMATES >>>>> BID	CONSTR. PHASE
(WHOLE UNIT)	+++++++						
UNIT OF USE	+++++++	++++++++					
AREA METHOD	+++++++	++++++++	++++++++	+++++++			
VOLUME METHOD		+++++++	++++++++	+++++++			
SYSTEMS METHOD			+++++++	++++++++	+++++++		
TRADE BREAKDOWN				+++++++	++++++++		
ENCLOSURE METHOD			++++++++	+++++++			
QUANTITY SURVEY-BASED METHODS IN-PLACE			+++++++	++++++++	+++++++		
MATERIAL/LABOR				++++++++	++++++++	++++++++	+++++++

FIGURE 3-12. Applicability of estimating methods during different project delivery phases.

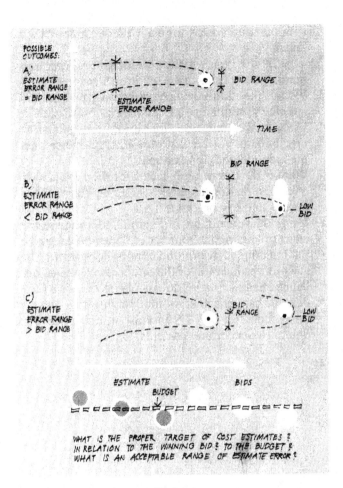

FIGURE 3-13.   The target of estimates.

Later on, a 10% or 5% difference may be too much for the client's comfort. However, the range of bids for a project quite often is much wider than that. The accuracy of an estimate can be known only after bids have been opened; but a client can get more reassurance from knowing that an estimator has had a consistent record of accuracy, with many estimates falling within 5% of the low bid, than from the precision of a given estimate in dollars and cents. However, this raises an interesting question with regard to reliability: what really is the target of the estimate—the low bid or the average of the bids submitted for a given project? Depending on the position we take on this issue, our acceptable range of error may vary widely.

WEIGHING COSTS AND BENEFITS OF ESTIMATING

The benefit of reducing our ignorance about the expected initial cost, that is, of trying to increase the precision of estimates in the hope of improving their accuracy, must be weighed against the cost of preparing an estimate. Given reliable data and conditions in which the data are applicable, precision is achieved

through more work, that is, simply by ascertaining more details about the building and applying their unit prices. Thus the cost is mainly the cost of the time spent on the estimate. Whereas unit-of-use and area method estimates can be established within a matter of hours for most types of projects, more detailed estimates will require more time—from a matter of days to several weeks, using conventional procedures. This is no more than we would expect, given what we know about the detail of available information about the building as the plans are refined.

It would seem that the rewards for achieving a more precise estimate (that is, one whose range of probable error is about 5 to 10 percentage points less than earlier rough estimates) are not worth the cost of having it done—even considering the risk of having to redesign if the bids come out too far over budget. The picture changes, however, when one takes into consideration not only the reduction of ignorance and risk but the use of an estimate to improve design decisions. Here, a different relationship emerges: the potential for savings, as well as the potential for making serious and costly errors, is greatest for the early design decisions, those made in the conceptual design and schematic design phases. This point is stressed in almost all textbooks on estimating and cost analysis. It is much more difficult, however, to draw the appropriate conclusions from it —that is, to increase the frequency and detail of early estimates, and to make good use of the results in producing improved design decisions. It is difficult because the methods available at early planning stages cannot distinguish very well between alternative design schemes for which only overall square footage is known.

Two main strategies can be seen for addressing this dilemma. One is that of increasing detail and precision early on. This is a main factor in alternative delivery methods such as design-build project delivery: bring the contractor into the design team from the beginning, and make sufficiently detailed decisions early enough that a firm price can be guaranteed to the owner even though not all aspects of the building have been worked out. This approach then allows an early start of construction and saves money mainly by saving time in the overall process. The second strategy would be to change the estimating methods used in early stages. First, advances in computer technology are making it possible to reduce the time and cost required for using estimating methods that previously were meaningful only at later stages—systems estimates and quantity-takeoff-based methods; these then can be applied much earlier to many "what-if" alternatives to find the best among several solutions. In particular, linking computer-aided design programs to up-to-date cost data bases will make this a realistic if somewhat roundabout possibility.

A second version of this strategy is to improve estimating methods to become sensitive to design variations in architectural form—bridging the current gap between the very rough and form-insensitive area method and the much more detailed quantity-takeoff-based methods. This is the thrust behind the recommendation to strengthen the enclosure method. Unfortunately, the format of cost data as currently collected and distributed does not favor this method. The cost data needed for this approach currently must be constructed from unit price data. However, it focuses much more directly on the crucial design decisions (form, height, etc.) and lends itself very well to the use of "quick-and-dirty" spreadsheet computer models that can easily be manipulated to find best solutions by approximation.

STUDY QUESTIONS

1. What is the difference between a cost estimate and the final cost accounting if both can be described with the same basic formula: Cost = Amount*Unit price?

2. What is the basis for the classification of cost estimating methods as discussed in this chapter?

3. What is the essential difference between owners' and architects' estimates and contractors' estimates?

4. Discuss the rationale for using an estimating method such as the enclosure method rather than the area method, for example.

5. What is the determining factor for the applicability of an estimating method at a given stage of the project delivery process?

6. Using the spreadsheet in Figure 3-10, examine the effect of varying the number of floors, the length of the building, its width, the floor-to-floor height, the glass-to-opaque ratio of the exterior wall, and so on. What changes in the design of the building (i.e., design changes only!) would you use in order to offset the effect of a 5% increase in unit prices for one of the enclosure elements?

4
Financing Construction Projects

INTRODUCTION: WHERE THE MONEY COMES FROM, WHERE THE MONEY GOES

The discussion in the preceding chapters mainly considered what costs arise in connection with construction projects. In other words, the concern was with where the money goes, and with estimating those costs to arrive at a sense of the total project budget.

Some of the costs were related to the process of borrowing needed funds, and those costs often are quite substantial. Also, they usually are not mere givens but can be manipulated and negotiated to a considerable extent. The magnitude of these cost items may explain why owners sometimes seem more interested in the specifics of the financing package for the project than in the architect's efforts to control construction costs. Unfortunately, the factor owners most often would like to see the architects control is not directly related to the quality of the architectural work itself; it is time.

It thus is necessary to take a closer look at the issue of where the money for construction projects comes from—first, in order to understand enough about it to make adequate provisions for it in the cost estimates and the budget proposals, and, second, to understand the relationships between financing provisions and design decisions. Are there any direct relationships? Are there "strings attached" to certain money sources or arrangements, and what design factors are affected by these strings?

This chapter will discuss the process and some fundamentals of construction financing. It will be helpful to have a good idea of the overall project delivery process (discussed in detail in Appendix 2) and the concepts of interest, inflation, the mechanics of time value

conversion of money, and so on (explained in Appendix 8).

MAIN PHASES IN THE FINANCING PROCESS

Figure 4-1 shows a simplified diagram of the three main sources of funds involved in a typical construction project. The costs of site acquisition, planning, and construction of the building are covered: (a) by the owner's own funds, the equity investment (down payment); (b) by money borrowed from one or several lending institutions; and (c) where applicable, by subsidies from government agencies, which may come in the form of loans or loan guarantees by the government, in the form of tax advantages, or (more rarely) in the form of outright funds supplied by the government.

Of course, this outline is not the whole story. It covers only the first step of getting the building built. Specifically, it addresses the first phase of financing a building, which is the phase of financing its construction. The main component at this phase, the construction loan (b), must be repaid in some way. If the project is sold immediately after completion, the sale proceeds would be used to pay off the loan. If the owner keeps the building, the construction loan must be converted into a long-term loan—the mortgage. These two steps occasionally may be fused into one, and the lender may be the same in both cases; but it is conceptually more clear to think of two distinct phases and two different lenders: one for the construction loan and one for the long-term or "permanent financing" loan or mortgage. The loan from the long-term mortgage is used to pay off the construction loan. For example, the buyer of a

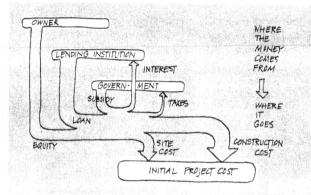

FIGURE 4-1.   The three main sources of funds for construction projects (simplified): construction financing.

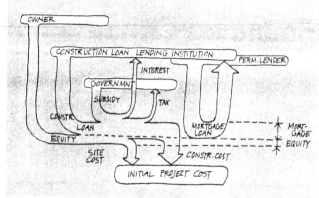

FIGURE 4-2.   Construction financing and "permanent" financing.

new building arranges for a mortgage to finance the purchase price for the project with bank A. The buyer then gives the money to the seller of the project, say a developer, who has borrowed the construction money from bank B, and uses part of it to repay this construction loan. Figure 4-2 shows this second phase.

For both types of loan arrangements there will be up-front financing charges ("points" or loan fees and closing costs). Government subsidies, in the form of either actual funds, loan guarantees, or tax advantages, also can apply to both kinds of loans. The difference lies in how the arrangements for repayment and especially how the interest for the borrowed money are treated. For the duration of the construction loan, only interest is being paid on a regular basis, and the entire

principal is repaid in one sum at the end; the long-term mortgage is repaid in equal installments that contain both interest and principal, so that the loan is paid off with the last payment.

Figure 4-3 includes, in addition to the prior two phases, the process of paying off the long-term mortgage. A new source of funds is introduced—the user, who may be the tenant paying rent for the use of the building or of a part of it. (Again, the user also may be the owner, but it is easier to think of the roles as separate in order to understand the process.) Some of the rent payments will be used to cover the costs of operating and maintaining the building, including taxes (this will be treated in more detail in future chapters). Another part will go toward repayment of the mortgage

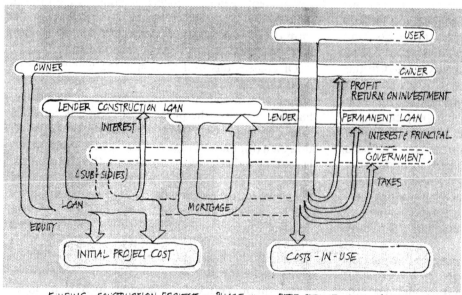

FIGURE 4-3.   Paying off the long-term loan (mortgage).

loan — "debt service." Finally, if the owner and the tenant are separate, the owner will expect part of the rent to repay his or her equity investment with interest, or profit.

All these payments occur with time delays; a loan is taken out to be repaid later. The amount of time involved in paying it back is one of the main factors involved in determining the cost of the loan and therefore will have to be studied closely.

Another point should be inferred from the diagrams. Besides the important distinction between sources of money (where it comes from) and its cost destinations (where it goes), it is of course a requirement that the amounts involved ultimately must correspond; the amounts from all sources must equal the sum of all combined costs. This requirement has led to some confusion between the concepts.

Finally, the diagrams should be seen as an invitation to think of the process not as a series of one-time transactions but as a continuous flow, to which ultimately the equivalency principle also must be applied. For example, funds used for government subsidies, if any, must either have been paid to the government at some time or be repaid to the government by someone. Thus, a building project might be visualized as a temporary trap in the overall continuous stream of money, trapped in order to produce value, or attract more money; but the money ultimately must return to the common stream, which is properly called "circulation" — it must circulate in order to keep working as a medium for the production and exchange of value.

CONSTRUCTION FINANCING

The Construction Loan

An owner or developer planning to build a project will arrange for financing its construction with a suitable bank, savings-and-loan, or other lender if the owner's own funds are insufficient to cover all the costs involved. It is rare nowadays to have an owner pay outright for an entire project. Typically, the construction loan will be a credit line in the amount of a certain percentage of the estimated project cost, the loan-to-value ratio, which will be determined in part by the amount of available equity on the one hand and by lender policy on the other. A frequent assumption is a loan-to-value ratio of 0.8 or 80% financing, which would require the owner to contribute 20% of the project cost as his or her equity in the project. In reality, an owner's or developer's "sunk costs" (the costs already incurred that cannot be recovered even if the project should not be realized), such as the cost of the previously purchased site and the site development costs, will be counted against this equity requirement so that the lender will supply all or almost all the actual funds

needed for construction. For many projects, 100% financing is no longer uncommon.

The lender will charge the borrower both for lending the money and for efforts involved in establishing the transaction. The latter is done in the form of "closing costs" — the various fees for services involved in setting up the loan: checking the applicant's credit, reviewing and processing the application, title checks, legal fees, surveys, documentary stamps, taxes, and so on — and "discount points," which may range from 1% to 4% or more of the amount borrowed. This amount is taken off the top of the amount actually disbursed; it constitutes one part of the cost of construction financing.

The other part is the interest on the loan. Interest is charged on the actual amount drawn against the credit line and usually is paid monthly. That is, as bills for materials and completed work on the project begin to come in, they are paid from the credit line. Interest then is charged month by month on the amount drawn, until the project is completed and the entire loan can be repaid. The actual amount of interest that must be paid thus depends both on the total construction time and on the pattern in which the bills for completed work come due. This pattern is called the "draw schedule."

For example, consider a project financed with a construction loan of $50,000 at 12% annual interest (that is, 1% monthly). The construction period is five months. If the draw schedule is simply $10,000 every month (Figure 4-4), the interest payments would be: 4*$100 = $400 for the first payment, 3*$100 = $300 for the second, 2*100 = $200 for the third, 1*$100 for the fourth, and none for the fifth, on the assumption that the project is sold and paid for and the construction loan repaid immediately upon completion. The interest total thus would be $1,000.

However, if the payments were $12,000 each for the first two months, $10,000 for the third, and $8,000 for the last two months, the interest payments would be as shown in the second column, and the total would be $1,120 for the five-month period, even though the total amount financed is the same. Now contractors, who most likely have themselves borrowed money for materials, equipment, and labor, naturally are interested in getting paid earlier rather than later. Some contractors therefore have shown a tendency to "front-load" their bids, meaning that work to be completed earlier will be relatively more expensive than work to be completed later. This strategy could become expensive for the owner, especially of larger projects, because of the interest on the construction loan, as shown.

Estimating Construction Loan Interest

How can the amount of interest on the construction loan be estimated at early stages of planning when the actual draw schedule obviously cannot be known? In

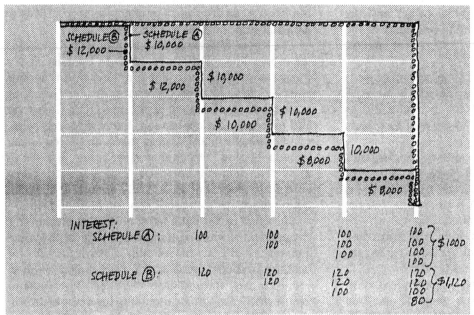

TWO DRAW SCHEDULES AND CORRESPONDING INTEREST PAYMENTS ON CONSTRUCTION LOAN

FIGURE 4-4.   Comparison of two draw schedules and corresponding construction loan interest payments.

practice, this is done not by assuming some hypothetical draw schedule, but by using the concept of the average outstanding balance of the loan. It is customary to assume that on the average, about 50% or 60% of the full loan would be drawn, and interest is calculated on this average balance for the entire period.

For the $50,000 project of the above example, using a 50% average outstanding loan percentage would result in an average balance of $25,000. One percent interest on this is $250 a month, paid for four months; the total amount of interest would be $1,000—the same amount as for the first draw schedule in Figure 4-4, and thus quite a good approximation. An average outstanding balance of 60% would result in a total interest charge of $1,200. Figure 4-5 shows a diagram for this approach.

Critical Variables: What to Watch for in Construction Loan Terms

Before we can answer the question of how construction loan terms relate to design decisions, we should look at the key variables that determine the amount of the charges involved. They include: the amount of the loan itself, and its relation to the borrower's equity contribution (if any), expressed through the loan-to-value ratio; the discount points; the interest rate and the method of calculating interest (e.g., monthly, weekly, daily); and, most important, the construction period during which interest must be paid.

For the example shown above, the total financing

costs for the construction loan with terms of 3% for discount points, a 12% annual rate of interest, a 50% average outstanding balance, and a construction period of five months, would be 0.03*$50,000 = $1,500 for the points, plus $1,000 for interest as calculated above, making a total of $2,500. This is 4.2% of the total project cost of $60,000 assuming a $10,000 equity contribution, corresponding to a loan-to-value ratio of 80%. Put in relation to the actual net construction costs: if we assume that the actual construction costs, for the sake of simplicity, are about $40/sf and the project has a size of about 1000 sf, the financing costs represent the equivalent of an increase of $2.50/sf or

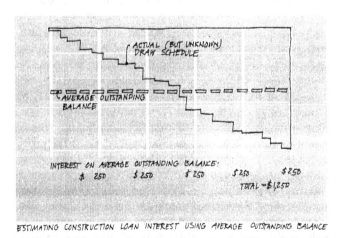

ESTIMATING CONSTRUCTION LOAN INTEREST USING AVERAGE OUTSTANDING BALANCE

FIGURE 4-5.   Estimating construction loan interest.

6.25% of the construction cost. Or, put differently, without the financing charges, the project could have included an additional 62.5 sf of floor area.

Construction Financing and Design Decisions

The designer's influence on the variables discussed normally will be minimal—except for the construction period, indirectly. But this influence is significant, especially for larger projects. The choice of structural system, construction methods and materials, and, most of all, simplicity or complexity of forms and details, combine to make a smooth, efficient, and fast construction process possible or very difficult to achieve for the people in charge of organizing the construction process itself. The industry's major efforts toward reducing the time needed for construction projects have focused on delivery process arrangements that bring the expertise of contractor and construction manager into the overall process as early as possible: fast-track and design-build. The idea is to bring this expertise to bear on the design process early on to produce a solution that then will enable the construction process manager effectively to apply modern construction planning, management, and scheduling techniques such as network scheduling.

"PERMANENT" FINANCING

The Permanent Financing Loan: Mortgage

The arrangements for financing a building project for long-term ownership after completion are somewhat different from those described above for the construction loan. To understand the difference, it helps to look at the situation of a building that is being planned and built by a developer for sale after completion—even if the actual arrangement in many cases is very different. The project initially must be financed by a construction loan taken out by the developer, who then is responsible for paying it back. The lender assumes a considerable risk in such situations because, after all, there is no real collateral for the loan yet, except for the site. Should the developer default on the interest payments during construction and have to give up the project, the lender would be left with an unfinished project, most likely already in some trouble due to interrupted work and resulting damage to previously completed parts and materials stored on site, litigation with subcontractors who had not performed adequately or who had not been paid, and so on. For this reason, lenders tend to charge higher interest rates for construction loans than for other loans. For security in assessing this risk, the lender looks at the reputation and the financial situation of the developer, besides the quality of the planned project itself. Finally, because the developer expects to

sell the project upon its completion, the loan terms reflect this expectation by providing for repayment of the principal in full at that time. Until then, only interest is charged.

The situation for the buyer of the completed building is quite different. First, there now is a completed building in a useful state to serve as the collateral for the loan. The lender thus assumes considerably less risk, which usually is reflected in lower interest rates for such loans than for the construction loan. Second, the lender is dealing with a different kind of borrower—for example, either a homeowner who intends to live in the house for a long period of time or an owner who expects to lease or rent the building to other tenants. In both cases, the loan cannot be repaid in full after a short time but must be paid off in small installments over a much longer period. This period—the mortgage term—in conventional fixed-rate mortgages, normally has been 20 or 30 years. For convenience the installments are arranged to be of equal size each period (i.e., the same amount every month or year) and include both interest and at least a small share of the principal. Thus, the ratio of interest to principal in the payments changes over time. The first payment will consist largely of interest (the interest is for the entire principal owed during that first period) plus a small amount of principal; the last payment consists mainly of principal and a small amount of interest, and, with that payment, the loan is repaid. For example, the last payment for a loan with a 10% interest rate will consist of 10% interest and 90% principal.

Obviously, these periodic payments must stand in some meaningful relationship to the owner-borrower's income, whether that income consists of rental payments from the building itself such as in an apartment building, or of the owner's personal income from a job or other sources, as in the case of a home. The mortgage payments must be "affordable"; they cannot exceed a certain fraction of the borrower's income. The lender thus will look not only at the quality and the value of the building but also at the borrower's income and other financial obligations in determining how much he or she can afford to borrow.

With interest rates rising to unprecedented heights for several years during the 1980s, the typical traditional mortgage became too expensive for many prospective homeowners, especially for young families. In response, the building and financing industry developed a bewildering variety of different (so-called creative financing) mortgage arrangements. Although some of these truly were innovative and creative solutions to the problem, others turned out to be quite dangerous to the unwary customer. Before we discuss these new forms of financing, and the question of how to estimate or calculate the resulting payments, it will be useful to look at the critical variables involved.

Critical Variables for Permanent Financing

In estimating the cost of financing, and assessing the terms of permanent financing arrangements, the following variables must be considered:

- The amount of the loan, called the principal.
- The equity or down payment (cash equity contribution).
- The loan-to-value ratio, which expresses the lender's policy or negotiated offer with respect to the relationships between the principal and the equity contribution.
- The length or maturity of the loan, also called the mortgage term.
- The interest rate.
- The size of annual or monthly payments or debt service payments, as determined by the principal, mortgage term, and interest rate together.
- Whether interest rates are fixed or variable (or adjustable). This feature often gives its name to the entire loan arrangement. In fixed-rate mortgages, the interest rate remains the same throughout the agreed-upon mortgage term. In variable-rate loans, the interest rate can change, and this then changes the annual or monthly debt service payments.
- If the interest rate is variable, how often it can change. Some arrangements provide for rate changes after every three-year or five-year period of the term. Others let the interest rate change more frequently, following general economic conditions or other considerations.
- For variable rate loans, whether there is a "cap" or limit to how much they might change. This is especially important for adjustable rate mortgages in which there is no fixed schedule of adjustments. Typically, one then would look for a provision limiting the percentage by which the rate can change over a year. For example, the lender might agree that the rate would not be raised by more than two percentage points per year.
- In any such variable-rate loans, what considerations determine how the rates will change. For example, they might be arranged to follow some index of average market interest rates; or they might be set at a certain number of percentage points above the Federal Reserve Board's discount rate (the interest rate at which the Federal Reserve lends money to the large banks), the average prime rate (the interest rate the large banks offer their best corporate customers), or some other prearranged pattern such as the anticipated earning pattern of the borrower.
- For some "new" financing packages, whether monthly payments have been kept low at the expense of the rate of equity buildup. The periodic payments may contain only a very small amount of retired principal, or none at all. This means that it will either take a very long time to pay off the loan, or the payments must be increased drastically later on. At the extreme, the arrangement may even provide for the monthly payment not to fully cover the interest in the early phases of the loan, which is equivalent to an increase of the amount owed, or "negative amortization."
- Especially for loans arranged at times of high interest rates, whether there is an opportunity for refinancing before the end of the mortgage term (that is, to pay off the loan and arrange for a new one), and what the conditions for such refinancing will be.
- Finally, whether there are substantial up-front charges. In addition to the routine fees and charges in connection with acquiring a site and setting up the deal (which would occur even if an owner paid the entire cost in full without borrowing—the closing costs), there will be "loan fees" or discount points that effectively reduce the amount of the principal that the borrower can take out even though interest is paid on the whole amount.

Mortgage Types: The Conventional Process and New Approaches

During the period of high interest rates in the early 1980s, according to some surveys, there were about 25 or 30 different types of mortgage packages being offered by lending institutions throughout the United States. It was difficult even for experts to stay abreast of this variety of possible arrangements and to evaluate them properly.

The following is a survey of some of the main types of mortgages. It is not meaningful to state in general which ones are more worthwhile than others; this must be judged against the particular borrower's financial situation and needs in combination with an assessment of the specific terms.

Fixed-rate mortgage. The "traditional" mortgage, with interest rates and therefore monthly payments remaining constant over the entire team—usually 20, 25, or 30 years. In times of rising or strongly changing market interest rates, banks are reluctant to offer these mortgages except at very high rates. For a while, they were not available at all. Figure 4-6 shows the typical pattern of declining balance and principal-to-interest ratio for traditional fixed-rate mortgages.

Flexible-rate mortgage. A mortgage whose interest rates can change, for example, according to some index (such as the Federal Home Loan Bank Board national average mortgage rate). They may or may not have "caps" or limits as to how often and how much the rate, or the payments, can change. (If the rate is not capped but the payments are, this may result in negative amortization, at least for a while.)

Balloon mortgage. A mortgage that resembles the traditional

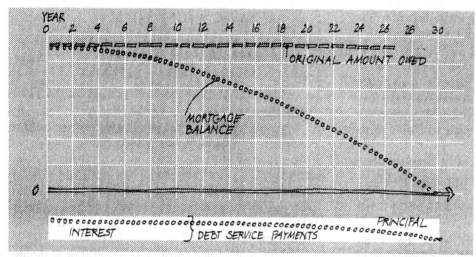

INTEREST TO PRINCIPAL RATIO & MORTGAGE BALANCE = TRADITIONAL MORTGAGE

FIGURE 4-6.   Balance and interest–principal ratio of a traditional fixed-rate mortgage.

fixed-rate loan with payments calculated as they would be for a long-term mortgage, except that the outstanding balance comes due in three or five years. It then must be paid in full, or the loan must be refinanced at then current rates. Some balloon mortgages have payments that cover interest only, in which case there is no equity buildup.

Graduated payment mortgage. Arranged for the "young family," a mortgage with initially low payments that then rise and level off later on, corresponding to the estimated pattern of increasing earning power of the family.

Growing equity mortgage. Mortgage in which the interest rates are fixed, but payments are arranged to rise on a schedule, to allow for a rapid payoff over a relatively short term (e.g., 15 years).

Shared appreciation mortgage (SAM). A mortgage that represents one of several attempts to make the better borrowing position of larger companies (developers or builders, who can obtain larger loans at lower rates) available to the individual home buyer. In return for below-market interest rates and payments, buyers agree to share with the lender a certain percentage of any appreciation of the house upon resale.

Figure 4-7 shows the effect of a mortgage with an initial period of negative amortization followed by a period of zero equity buildup, on the payments and the outstanding balance.

An example of calculation of the debt service charges for a traditional fixed rate mortgage may clarify how the various variables relate to one another. We assume a project with a cost (or purchase price) of

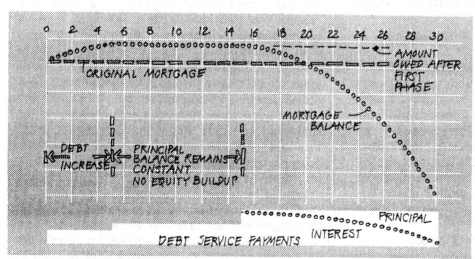

A MORTGAGE WITH VARIABLE PAYMENTS; RESULTING IN NEGATIVE AMORTIZATION

FIGURE 4-7.   A mortgage with negative amortization followed by a period of zero equity buildup.

$1,000,000, for which the lending institution is willing to offer a loan-to-value ratio lvr, of 0.85 (in this case, it is actually a loan-to-cost ratio) or 85% financing. This means that the loan will be $850,000. The interest rate is 10% annually and the mortgage term 30 years. There will be discount points of 2.5%. The resulting loan fee is therefore 0.025*$850,000 = $21,250. This will be subtracted from the actual amount made available to the borrower.

The periodic mortgage payment or debt service is calculated by means of the capital recovery factor:

$$CRV = (MINT*(1 + MINT)^m)/(1 + MINT)^m - 1)$$

(calculating the annuity equivalent, given the present value of the loan; see Appendix 8) for an annual interest rate of 10% and 30 years. If we wish to find the annual debt service payment, this factor would be:

$$(0.1*(1 + 0.1)^{30})/((1 + 0.1)^{30} - 1) = 0.106079$$

and the annual payment is:

ADS (annual debt service) =
$$0.106079*\$850,000 = \$90,167.36$$

Alternatively, if we wish to find the monthly payment, m and MINT first must both be converted to a monthly basis: $m + 12*30 = 360$, and $MINT = 0.1/12 = 0.00833$. Then:

$$CRF = (0.00833*(1.00833)^{360})/ (1.00833)^{360} - 1) = 0.008775$$

and:

MDS (Monthly Debt Service) =
$$0.008775*\$850,000 = \$7,459.36$$

(Note that this result is slightly different from that obtained by simply dividing ADS by 12.)

For various purposes such as early repayment or tax deduction of interest for home mortgages, it often is necessary to calculate the amount of interest contained in a specific debt service payment, the principal contained in the same payment, and the remaining outstanding balance at a given point in time. There are two ways of doing such calculations. The first involves step-by-step or period-by-period calculations, starting with the initial loan balance before the first payment:

Mortgage balance:
$$MORTBAL_0 = LOAN = \$850,000$$

Debt Service:
$$ADS_1 = \$90,167.36$$

Interest first payment:

$$INTAMT_1 = MINT*LOAN$$
$$= 0.1*\$850,000 = \$85,000$$

Principal retired in first payment:
$$PRRET_1 = ADS - INTAMNT_1$$
$$= \$5,167.36$$

Mortgage balance, after first payment:
$$MORTBAL_1 = MORTBAL_0 -$$
$$PRRET_1 = \$844,832.64$$

and so on. This approach may look cumbersome, but it is a convenient method when used with spreadsheets to print out the entire payment schedule with year-by-year interest, principal, and remaining balance, using the "copy down" command or equivalent.

The other approach consists of using the appropriate series-payment/present worth factor for any desired period i:

$$MORTBAL_i = DS*((1 + MINT)^{m-i} - 1)/ MINT*(1 + MINT)^{m-i}$$

for example, the balance for year 20 would be:

$$MORTBAL_{20} = ADS*((1.1)^{10} - 1)/0.1*(1.1)^{10}$$
$$= \$90,167.36*0.614457 = \$554,039.39$$

Interest and principal are calculated the same way as shown above, by multiplying last year's balance with the interest rate to find the amount of interest paid that year, and subtracting this amount from the debt service payment to find the amount of principal retired that year.

Permanent Financing and Design Decisions

It may seem from the example and the list of critical variables that there is little if any connection between these financing terms and architectural design considerations that might influence them. In reality, there are quite a few such connections, which often are somewhat roundabout and come in the form of "strings attached" to the availability of or eligibility for certain kinds of mortgages. In particular, government-subsidized financing for housing, for example, is contingent upon the design's conforming to many standards and constraints, including overall unit size, room sizes, dimensions, type and extent of insulation, sizes of window openings in relation to floor area, and so on. Thus, in order to ensure that a project will qualify for a desired form of government financing program, say, the Federal Home Administration or similar programs, the designer will be well advised to become quite familiar with the pertinent design guidelines for the program in question before starting the work. Failure to do so may mean costly delays and unnecessary work, with repeated revisions to accommodate the necessary changes.

Private lenders may have less stringent requirements for specific dimensional constraints but will instead look at the design even of a house built for a specific individual from the point of view of how easy or difficult it would be to sell on the open market, should the owner default on the mortgage payments. They will balk at projects that are so unusual, unique, and different from common expectations that they would be difficult to sell to another buyer. Indirectly, this results in a subtle form of standardization.

FINANCING PUBLIC PROJECTS

Public projects often have some unusual implications for the practitioner with regard to their funding and its impact upon the planning and design process, starting with the sources and the forms of funding.

Like private owners and companies, public institutions such as government agencies usually must borrow money for large construction projects. Though it would be possible for large institutions to pay for a project outright, this would not be good policy if it led to big fluctuations in the annual budget. Financing in any form is a way to spread out the cost of large construction projects over time, to match the pattern of income over time, whether that income is from rents or taxes. The large scale of such projects, for example, highway projects, makes it impractical to finance them through banks and invites a different mode of financing, that of bond issues. The agency or the government will sell shares of the bond issue to private investors on the financial markets, just like stocks. Because these bonds are guaranteed by the taxing power of the respective government, they are considered much safer investments than regular stocks, and therefore are offered at lower interest rates than the dividends of stocks or certificates of deposit of the same size. They often are combined with tax advantages (e.g., that no taxes are paid on the interest earned on such bonds). Like stocks, such bonds can be traded, and the selling prices at which they trade are indicators of the degree of confidence the market has in the particular government and its tax base, which in turn determines the interest rates at which they are offered.

There are two main kinds of bonds: general bonds that are not specifically linked to a particular project, and are guaranteed by the government's general taxing power, and special bonds intended, for example, for a specific highway or bridge project, which pay dividends according to the success of the project, say, the highway tolls generated. Obviously, special bonds are much closer in nature to stocks offered by private companies, carry higher risk, and therefore are offered at higher interest rates than general bonds.

The considerations involved in determining the source of funding for public projects are usually not of so much concern to the architect as the complications arising from the process of getting such projects approved through the chain of review agencies involved. The number of such agencies can be considerable. This will not only influence the delivery process; there usually are much tighter "strings" attached to such funding, which have direct design implications. Often, these design constraints come in the form of design or planning criteria issued by the departments or agencies responsible for that particular class of projects or building type. Examples are the design criteria developed for educational facilities, university systems, health care facilities, corporate facilities such as hotel chains and restaurant (e.g., fast food) chains, and projects in certain environments such as coastal areas. It is advisable to carefully study the capital funding procedures for every public project in which one gets involved, and the relationship of these procedures to the corresponding design criteria and constraints. Although it is not meaningful to go into any such specific process in detail—they will differ with the locality, state, building type, and level of government—it is important to understand the basic logic and rationale underlying most of the variations of these regulations.

A single institutional or public project, such as a new academic building for a university, usually will be one of many projects funded by the responsible agency at the same time. Because funds for such projects are limited, many projects are competing for available funds. Thus it is necessary to establish some guidelines for determining their respective priority, beginning with the question of whether, when, and under what conditions a new project actually is needed. To be able to compare applications for projects to be funded, rules are developed that describe, in broad outline or sometimes in considerable detail, program requirements and even design provisions for projects of the type in question. This allows reviewers to match projects against needs criteria. An example is the "Certificate of Need" that must be established in order to gain approval for new health care facilities.

Thus, there often is a whole detailed application process to be successfully completed in order merely to gain approval to begin site selection, planning, and designing for a new facility, and to seek funding for it. In the course of this application and approval process, important program and design features of the new project can become part of the conditions for the planning approval. These features sometimes, but not always, are linked to the design criteria or planning guidelines developed by the agency or agencies involved. The approval in many cases also carries a stipulation for the project budget. Approval to proceed with the planning for the facility must be obtained before the delivery process can begin that an architect accustomed to private projects would expect. However, this process too will be different from that for a private project in many respects. Not only will many program and design provisions be expected to be adhered to as a matter of course, by virtue of the guidelines adopted by the agencies involved, but every document constituting the basis for the next "go-ahead" decision must be reviewed and checked for compliance with those guidelines.

The rationale for the design guidelines and criteria used for these purposes varies. Ostensibly, there is an issue of fairness and equal treatment for all projects of the same kind under the jurisdiction of the same agency. There is no reason why faculty offices or classrooms for one department should be funded and designed twice as lavishly as those for another department across campus, or on another campus in the same university system. In other cases, quality and safety concerns may be the driving force behind the effort to standardize. To ensure that all projects live up to the same level of expectations, it is essential to have some standard, and an orderly procedure for checking applications for adherence to that standard.

Two kinds of such standards can be distinguished. One essentially prescribes design solutions, specific materials, or products; examples are the product specifications that still dominate most construction specification documents. The other first was developed systematically in connection with industrialized building efforts such as the SCSD school systems, and aimed to open up the market for new (industrialized) products that could not be specified as such without falling into the extreme of product specification—brand name specification. This is performance specification, describing what a material or building component should do or achieve, so that many different products might compete for the intended use. The concept is, of course, applicable at many levels, from program to construction document and from urban design to individual details. A case could be made that all design guidelines and criteria should be written as performance standards. The difficulty lies in describing precisely what the desired performance should be, expressing this in objectively measurable terms, and carrying out the required tests and measurements.

Whichever approach is predominant in a given situation, the resulting second review process for compliance with guidelines, standards, and criteria sometimes can constitute a formidable array of obstacles to a smooth design delivery process. The architect will be well advised to become thoroughly familiar with the set of such regulations applicable to the respective new project. The aim should be to design the project to meet the stipulated criteria the first time around, rather than to revise and resubmit the plans in response to objections, which could have a disastrous effect on project delivery time and thus on the overall cost.

In the following chapters, the examples and the general discussion will refer to the comparably simple case of conventional financing for private, not public, projects.

STUDY QUESTIONS

1. A prospective homeowner can count on an annual income of $30,000. The lending institution stipulates that the annual debt service payments (excluding utilities and other ongoing costs-in-use) not exceed 20% of the total household income. Assuming further a loan-to-value ratio of 0.80, a loan interest rate of 10% for a term of 30 years, and that the client has a sum of $7,000 of savings to use as the equity contribution (down payment), how much can the owner afford to borrow, and how much can the project cost?

2. Look at your latest design project, and identify some design constraints that can be traced to the question of the funding source and its specific requirements. What might the design solution have looked like if those requirements had not been in force?

3. Consider the case of a 100,000 sf five-story building for which the designer must choose between cast-in-place concrete slabs for the five floors and steel construction. The concrete slab construction will cost $6/sf of TFA, whereas the steel construction will cost $7/sf. Thus, the difference in cost is $100,000. However, the concrete slab method will add one week per floor to the construction period plus three weeks of overall delays because some subcontractors must leave and return to the site several times to complete their work.

   (a) At an estimated 50% average outstanding balance percentage, a six-month construction period for the steel building, and a 12% interest rate for the 80% financed construction loan, is the concrete solution still preferable?

   (b) How much of a change in construction time would it take for the two methods to be equal in economic (i.e., construction and financing cost) terms?

   (c) How much would the construction prices have to change for the two methods to become equal in this case?

   (d) How would the solutions compare if one or the other solution (check both) also were to result in a 60% average outstanding balance percentage?

4. Calculate the required annual (monthly) mortgage payment for a $120,000 home with a 75% (80%) loan-to-value ratio, a 9.5% (10.5%) interest rate, fixed for a mortgage term of 30 (20) years. What is the interest paid in year five (15) and the corresponding principal retired during that year? What is the mortgage balance at the end of year five (15, 20)?

5
The Future Performance of Buildings: Cost-in-Use; Life Cycle Cost

OVERVIEW

Architects traditionally have been concerned with correctly estimating and controlling the initial cost of buildings. However, the initial cost represents only a fraction of the long-term costs associated with owning and operating a building over its life span. In recent years, more attention has been paid to life cycle cost analysis, especially since the energy crises of the 1970s and 1980s. Energy costs do play a major role in the long-term costs-in-use, but they are only one of many cost factors that must be considered and understood if the architect is to make meaningful design decisions. Of course, both initial and future costs must be weighed against the benefits of the building: are they worth the costs? The benefits are expected to occur in the future; even if the building is sold immediately after completion, there is a delay between the initial investment and the accrual of benefits that must be taken into account.

This chapter provides an overview of the different kinds of costs that must be included in an analysis of long-term and life cycle cost for building projects, as well as estimating methods for such costs. Chapter 6 examines the benefits (both monetary and intangible) of buildings, and Chapter 7 addresses the task of relating costs and benefits to one another to form comprehensive measures of economic project performance. Chapter 8 discusses some common techniques for economic performance analysis that use one or several of those performance measures.

First, we must gain an understanding of what costs and benefits items will occur in the long run, and how they relate to one another. Figure 5-1 shows a simplified cash flow diagram with typical main categories of

monetary costs and benefits: initial cost, construction loan, mortgage loan, rental income and sale proceeds, debt service, operating expenses, maintenance, repair and replacement costs, taxes, and sale cost.

Not shown are the nonmonetary benefits or use values for which people are willing to pay the rent or purchase price of a building. Nor, for that matter, does the diagram show the intangible costs and disbenefits associated with the building project, such as the loss of vegetation on the site, traffic increase resulting in increased pollution, noise, runoff effects, and changes in image and quality of the neighborhood that may be caused or accelerated by the project (which may be both positive and negative, to be sure, and also may be regarded as positive or negative by different parties affected by the change). Figure 5-2 adds these intangible costs and benefits to the picture in a general way.

Future Cost Elements

Figure 5-3 shows some of the main categories of monetary costs associated with owning and/or operating a building year by year. It does not say anything about how these costs (or expenses) are paid for. They might be borne by the owner of a building or by a tenant; that is a matter to be arranged between owner and tenant in the lease or rental agreement. These costs will be discussed in more detail in the following section.

ESTIMATING FUTURE COSTS

Operating Expenses

A number of costs that are necessary to operate a building, maintain it, heat, cool, and service it, and so on,

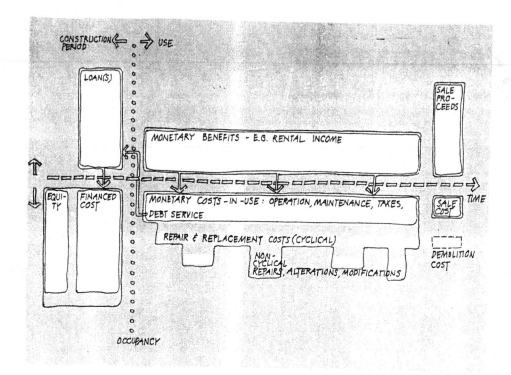

FIGURE 5-1.   Overview of monetary costs and benefits of building projects.

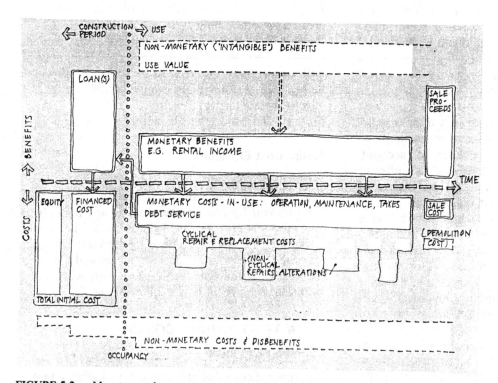

FIGURE 5-2.   Monetary and nonmonetary costs and benefits.

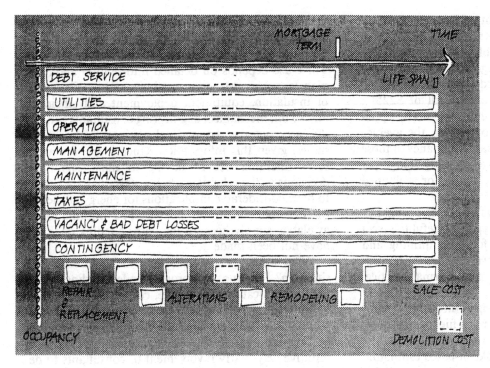

FIGURE 5-3.   Elements of building costs-in-use.

are lumped together as operating costs or operating expenses. What these costs are called and how they are grouped often depends on the building type and such things as the arrangements in the lease concerning who pays for which items. For example, in a commercial office or an apartment building, the tenant usually is billed directly by the utility company for electricity, water, gas, and so on, used inside the leased space. But there also will be utility charges for the common areas in the building, which are paid by the owner and recovered in the rent. Such costs also include, for example, fuel costs not covered under utilities but provided by the owner/operator of a building (oil, coal, gas); emergency equipment costs; the costs of operating elevators and escalators and of heating and air-conditioning public spaces in the building; and the salaries of operating personnel.

Operating costs commonly are expressed as dollars per square foot of building area per year, but they also can be given in terms of dollars per unit of use (work station, hospital bed) or per unit of surface area of building enclosure.

If an owner is also the user of a building, and the use involves the operation of equipment integrated into the building (fume exhaust hoods in laboratories, cranes conveying equipment, etc.), the task of sorting out operational costs for the building will become rather complex, and they will be difficult to separate from the costs of using the space, for example, production costs of goods manufactured in the building.

See Appendix 13 for the different types of leases, distinguished according to what costs are covered by the rental rates as opposed to being paid directly by the tenant. The following discussion of the nature of these costs and how they can be estimated will ignore the issue of who pays for what.

Utilities

Utility costs cover services provided by the utility company: water, electricity, gas, sewage tax, garbage collection, and so on. Utility costs for the general operation (e.g., outdoor and hallway lighting, elevators, etc.) of a multi-user rental building must be distinguished from the utility costs that can be attributed to specific rented space and therefore charged directly to tenants. Rates vary both with location (municipalities have different utility tax rates) and overtime; in many places, different rates apply to small-scale (residential) and heavy (e.g., industrial) users. Rates can be obtained from the local utility office. They are expressed in terms of the metered units for some items: kilowatt-hours, gallons of water, cubic feet of gas, gallons of sewage, and so forth. For other items such as garbage collection, flat fees are charged. In order to make useful estimates of what the utility usage, and thus the costs, will be for a planned project, assumptions must be made about average usage, either per person per time unit and number of users occupying the building or per square foot of floor area per time unit.

For example, the average utility bill for single-family detached dwellings in a neighborhood may be $180 per month. Given the average size of the homes, say 1200 sf, one might arrive at an average rate per square foot per month of $0.15; for a planned new home of 1500 sf, one then would estimate a monthly bill of $225 — assuming that general construction methods, appliances, and so on, are similar to those in the average home. Or, knowing that the average occupancy of those homes is 2.5 persons, the average monthly charge per person would be $72, and the estimate for a planned four-person household would be $288.

Such estimates based on square foot or occupancy averages for the given building type are useful to give a rough idea of expected costs, for programming and feasibility analysis purposes. However, once the considered design solutions deviate from the normal practice, or efforts are made to achieve, for example, greater energy efficiency through design measures, better methods for estimating such items as fuel or electricity use are necessary.

Again, as with the quantity-takeoff-based estimating methods for initial construction cost, there are reliable detailed estimating methods — for example, for energy requirements for artificial lighting, for household water heating, for heating/cooling/ventilation, and so on. Their use depends on detailed information about the plan layout, construction methods and details, extent of insulation, equipment deployment and efficiency, and so on; mechanical engineering consultants can easily carry out the required calculations for such estimates, given this information. Once more, the architect's dilemma lies in having to make significant decisions at the schematic design stage that are certain to have considerable implications for energy efficiency, for example, before all this detailed information is available, simply because the corresponding decisions have not been made.

What strategies are available to the architect in this situation? For one, a thorough understanding of the energy implications of climate, construction methods, building orientation, and the mechanical equipment used in the building can help the designer avoid blatantly wrong moves, by using this knowledge in an intuitive way. Early involvement of consultants to determine the general design strategy also can help if the

EXAMPLE - INITIAL AND ANNUAL COSTS-IN-USE; ENCLOSURE METHOD

ASSUMPTIONS

Program/given			unit	formula
Net leasable area	NLA	80,000.00	sf	
Site area	SA	200,000.00	sf	
Assumptions				
Corridor width	CW	6.00	f	
Exterior wall thickn.	WTH	0.50	f	
Floor-to-floor height	FH	10.00	f	
Office depth	OFD	20.00	f	
Staircase width	STW	10.00	f	
Elevator lobby width	ELOBW	10.00	f	
Parking requirement std.	PARQ	200.00	sf NLA / prk sp.	
Parking spaces needed	PARSP	400.00	NLA/PARQ	
Parking area/space	PARSPQ	280.00	sf/space incl. drive	
Parking/drive area needed	PARKAR	112,000.00 sf	PARSP*PARSPQ	
Landscaping requirement	LSRQ	0.10	% of SA	
Landscaped area required	LSARQ	20,000.00	sf	LSRQ*SA
Number of floors	NOF	5.00	#	
Leasable area / floor	LAF	16,000.00	sf	NLA/NOF
Office width (total)	OFW	400.00	f	LAF/(2*OFD)
Building length	BL	421.00	f	OFW+STW+2*WTH+ELOBW
Building width	BW	47.00	f	2*(OFD+WTH)+CW
Building footprint	BFP	19,787.00	sf	BL*BW
Total Floor Area	TFA	98,935.00	sf	BFP*NOF
Actual landscaped area	ALSAR	68,213.00	sf	SA-BFP-PARKAR
Exterior wall area	EWALLAR	46800.00	sf	NOF*FH*2*(BL+BW)
Glazed/opaque ratio	GOR	0.40	%of EWALLAR	
Glazed area (windows)	GLAR	18720.00	sf	EWALLAR*GOR
Opaque wall area	OQAR	28080.00	sf	EWALLAR-GLAR
Office module width	OFMW	20.00	f	
Number of off. partitions	OFPART	20.00	f	OFW/OFMW
Interior wall area	IWALLAR	24210.00	sf	NOF*((OFPART*FH*OFD)+(2*BL))
No. elevators per flr.	NEL	2.00	#	
No. elev. landgs	NELNDS	10.00	#	NEL*NOF

Data assumptions

Variable	Symbol	Value	unit
A/E Fee percentage	AEFEEPC	10%	%of CCST
Development Price	DEVPRC	1	$/sf of SA
Legal &Acc. fee perc.	LEGFEEPC	3%	% of CCST
Cnstr. contingency %	CCONTPC	0.05	%of CCST
Permit percentage	PRMPC	0.02	%of CCST
Real est tax mills	MILLS	0.025	% of SCST
Construction period	CP	15	months
Constr.loan-value ratio	CLVR	80%	of SUBTOT
C. loan average outst. balance percent.	AVOBPC	60%	of CLOAN
Constr. loan interest	CLINT	12%	of CLOAN/yr
Discount points	DSCPTSC	0.03	%ofCLOAN
	DSCPTSM	0.02	%ofMLOAN
Mortg. interest rt	MLINT	0.095	%of MLOAN/yr
Mortgage term	M	25	years
Mortg. ln/val rat.	MLVR	85%	of TIPRJCST
Equity discount rate	DISC	0.12	%annual

FIGURE 5-4.   Spreadsheet for an enclosure model estimate of costs-in-use. Part A. Data assumptions.

consultants are willing to look at alternative options with incomplete information. Architects sometimes complain that their consultants do not like this imprecise and messy work in the early stages, preferring to take a given solution as it is and to calculate the requirements for it. Another possibility would be to apply an approach equivalent to that of the enclosure method for estimating initial cost—in fact, to expand the enclosure method into the realm of future costs. An example of a general study of this kind was presented in Kemper's *Architectural Handbook*, Chapters 3 through 8. It would involve, for example, distinguishing enclosure elements with different heat transfer characteristics so that future cost implications of schematic design alternatives could be compared by means of approximate "quick-and-dirty" models. A rough spreadsheet version of this approach is shown in Figure 5-4.

The enclosure method model spreadsheet for the schematic office building discussed in Chapter 2 is expanded in Figure 5-4 to show approximate energy costs for heating and cooling for each enclosure element—mainly different types of exterior wall construction and materials. The spreadsheet will calculate the annual costs of schematic design alternatives that result from changing the dimensions of the building. Provisions for the effect of building orientation can easily be added by distinguishing surfaces with different exposures and adjusting the annual cost figures to reflect, for example, added energy requirements for removing heat gained through east/west elevations or for compensating for heat loss to northern exposure (in the northern hemisphere).

The problem with this approach, as in the initial cost enclosure method, is that reliable data in the form needed are not readily available. They must be reconstructed from more detailed calculations of specific materials and assemblies, and approximate averages used to reflect the fact that specific decisions about materials and so forth may not have been made at the schematic design stage. But even with rough approximations it is possible to get a good sense of the relative cost effects of schematic design trade-off decisions. The model can include cost assumptions for other kinds of in-use costs (e.g., maintenance costs), discussed below on a per-square-foot basis. If accurate cost data are not available, indicators (see Appendix 9) such as the perimeter–area ratio or surface–volume ratio for a building can provide guidance for the design effort.

INITIAL				Initial price		Initial		ANNUAL Unit cost/yr		Repair &	Total
Area / element	Symbol	Value	Unit	Symbols	Value	cost (...CST)	Formula	Operation	Mainten	Replacemnt	annual
Site area	SA	200,000.00	sf	SPRC	$5.00	$1,000,000.00	SPRC*SA	$0.05	$0.02		$0.07
Landscaped area	ALSAR	68,213.00	sf	LSPRC	$5.00	$341,065.00	ALSAR*LSPRC	$0.02	$0.50	$0.20	$0.72
Parking &driveways	PARKAR	112,000.00	sf	PARPRC	$4.50	$504,000.00	PARPRC*PARKAR	$0.10	$0.35	$0.30	$0.75
Serviced space (floor area)	SSP	98,935.00	sf	SSPRC	$37.50	$3,710,062.50	SSPRC*SSP	$2.00	$0.80	$1.00	$3.80
Foundation/ground contact	FNDAR	19,787.00	sf	FNDPRC	$5.00	$98,935.00	FNDAR*FNDPRC	$0.00	$0.01	$0.10	$0.11
Roof area	RFAR	19,787.00	sf	RFPRC	$7.00	$138,509.00	RFAR*RFPRC	$0.30	$0.03	$0.10	$0.43
Glazed area (windows)	GLAR	18,720.00	sf	GLPRC	$21.00	$393,120.00	GLAR*GLPRC	$2.00	$0.20	$0.15	$2.35
Opaque extr. wall area	OQAR	28,080.00	sf	OQPRC	$16.00	$449,280.00	OQAR*OQPRC	$0.75	$0.10	$0.15	$1.00
Interior wall area	IWALLAR	24,210.00	sf	IWPRC	$2.50	$60,525.00	IWALLAR*IWPRC	$0.00	$0.10	$0.05	$0.15
Elevators	NEL	2	#	ELPRC	42000	$84,000.00	NEL*ELPRC	$200.00	$600.00	$1,500.00	$2,300.00
Additional landings	ADDLND	6	#	ALNDPRC	$7,000.00	$42,000.00	ALNDPRC*ADDLND	$50.00	$60.00	$100.00	$210.00
							Annual subt				$2,519.38
Subtotal building	BLDCST					$4,976,431.50	Annual/sf TFA				$9.41
Construction cost	CCST					$5,821,496.50					
Constr. contingency	CCONT					$291,074.83					
Subtotal	SUBTOT					$7,112,571.33	SCST+CCST+CCONT				
A/E Fees	AEFEES					$711,257.13	SUBTOT*AEFEEPC				
Permit costs	PRMCST					$116,429.93	PRMPC*CCST				
Real est.taxes	PROPTX					$25,000.00	MILLS*SCST				
Carrying charges						$1,000.00					
Project Cost	PRJCST					$7,966,258.39					
Constr.ln interest amnt	CLINTAMNT			$512,105.14			SUBTOT*CLVR*AVOBPC*CP/12*CLINT				
Constr. disc.points	CLOANFEE			$170,701.71			SUBTOT*CLVR*DSCPTSC				
Constr. financing cost	CFINCST					$682,806.85					
Tot. Init. Proj. Cost	TIPRJCST					$8,649,065.23					
Mortgage Loan	MLOAN			$7,351,705.45			TIPRJCST*MLVR	Debt service	DS		$778,982.24
								Total Annual cost until yr 30 w/o equity			$1,709,502.54
Cash equity contribution	CEQC			$1,297,359.79			TIPRJCST-MLOAN	Equity Uniform Annual Equivalent			$165,413.33
								Total annual incl. equity			$1,874,915.88

FIGURE 5-4.   Spreadsheet for costs-in-use. Part B. Calculations.

Maintenance Costs

This item includes janitorial services and cleaning crews, window washers, painting, wall finish upkeep, fixture cleaning, relampers, monitoring and servicing of mechanical equipment, and so on. Data for maintenance costs are available in various formats from different sources. One format would be records of the client's own organization if it is a corporation or institution that already owns and operates a number of facilities of the same type. Associations and government departments responsible for certain building types also keep records of such data, such as those for shopping malls, hospitals, educational facilities, hotels and motels, and so on. Data of this kind should be viewed and used with considerable caution because they are subject to manipulation by the owner. Also, policies differ widely as to what level of maintenance is adequate and appropriate. Should the building be kept at its initial value and utility (e.g., income-producing ability) for a time that in theory could be indefinite? Or should it meet a rising standard, being adapted to new technology and quality expectations over time? Or should the building simply be kept in acceptable working condition up to the end of its expected life span, allowing for a decreasing quality standard with age and obsolescence? The various trade-off decisions involving maintenance costs are well known and recurring: Should a more expensive material be used in expectation of lower maintenance costs over the years? Even if it must be replaced earlier? Should more insulation be added to the exterior enclosure (insulation in walls, insulating glass in windows, etc.) in the hope of reducing the continuing fuel costs, which are expected to rise even faster in the future?

Grounds maintenance refers to the costs of cleaning the outdoor premises, such as lawn maintenance, trimming, landscaping services, and trash collection.

Administrative Costs

Management. The costs of managing the building can vary considerably, depending on the type of building, the tenant turnover rate, the type of lease or rental agreement, the need for advertising, whether the owner or the owner company performs these duties in-house or engages outside help, and so on. They would be expressed, for example, in terms of dollars per square foot per year, or dollars per unit (say, apartment, hotel room, etc.) per year. There is little evidence for a straightforward relationship between design features and management costs that might or should influence design. Accounting and legal costs might be listed under this heading or separately.

Security Costs. These costs can include the operation and maintenance, for example, of electronic security systems and/or the services of night security guards or security firms. Again, cost assumptions would be expressed on a dollar-per-square-foot basis. While the relationship between design and the likelihood of crime in a building or neighborhood has been pointed out forcefully by authors such as Newman (see his *Defensible Space*) the impact of design decisions on the required level of security expenditure is much less clear. A great deal also depends on the location of the project.

Insurance. The cost of insurance coverage for a given building project depends very much on the nature and the use of the building, the kinds of hazards covered, the level of risk associated with each, and the level of competition among insurance companies offering a particular kind of coverage. Hazards and risk depend in turn on construction methods and materials, as well as the location (e.g., distance from fire stations and fire hydrants; also there are maps showing the expected frequency and levels of floods for the purpose of flood insurance, and insurance companies keep records of the number of burglaries or fires in different neighborhoods and use them to determine risks and premiums). It thus is difficult to give general reliable averages for the purpose of long-term insurance cost estimation. It will be necessary to obtain data from nearby similar buildings, or to conduct inquiries with insurance representatives for each particular project.

Telephone and Other Communications Charges. Again, a distinction must be made between the general service for a building (public phones, basic service and hookups provided, service to the manager's office, etc.) and the specific user (tenant) charges that can be metered and charged to individual user accounts.

Debt Service

The term "debt service" refers to the periodic (monthly or annual) payments required to pay back the mortgage loan for the construction or acquisition of a building. These payments usually are arranged to be constant (the same amount every year or month). Conceptually they are kept separate from other annual expenses, which will vary with inflation or other price fluctuations. There may be several loans or mortgages involved, especially in larger projects. Debt service payments most likely constitute a major part of the costs to be paid in connection with the use of a building. They are the initial costs (plus financing charges) spread out over time.

Debt service payments are calculated by means of the Capital Recovery Factor, CRF (see Appendix 8), or the mortgage constant, $(\text{MINT}(1 + \text{MINT})^m)/(((1 + \text{MINT})^m) - 1)$, where MINT stands for the periodic mortgage interest rate, and m is the mortgage term (number of periods). If m is expressed in years, then

MINT must be expressed as an annual rate; if a monthly payment is to be calculated, MINT must be expressed as a monthly rate, and m must be expressed in months. The factor can be obtained from mortgage tables or calculated by the above formula.

To obtain the periodic payment, CRF must be multiplied by the loan amount. For example, for a 30-year conventional fixed-rate mortgage of $100,000 with an annual interest rate of 8%, the CRF for the annual payment will be 0.08883 (from tables), and the annual debt service payment is $8,883. For the monthly payment, m will be 360, MINT is 0.006667 and the corresponding CRF is 0.007338, which will produce a monthly payment of $733.80. (Note that this result is slightly different from what we would get by simply dividing the annual $8,883 by 12; the difference is due both to the rounding errors of multiple exponentiation in the calculating equipment and to the fact of multiple exponentiation itself. Thus it makes a difference whether compounding is done on an annual, quarterly, monthly, daily, or continuous basis.)

If a different form of financing is chosen, for example, an annually adjustable variable-rate mortgage, the annual or monthly payments will not be fixed for more than one year; it then becomes correspondingly more difficult to estimate what the future payments will be. It might be worthwhile to calculate the required payments for the worst case—that each year the rate will be increased by the maximum stipulated in the arrangement, say one percentage point per year, up to a maximum rate of 18% annually. For the $100,000 mortgage at an initial rate of 8%, the annual payment would be $8,883 for the first year, $9,734 for the second year, $10,608 for the third year, $11,502 for the fourth year, $12,414 in year five, $13,341 in year six, $14,280 in year seven, $15,230 in year eight, $16,189 in year nine, $17,155 in year ten, and $18,126 from year eleven onward. Whereas the total payments for the fixed-rate mortgage amount to $8,883 \times 30 =$ $266,490, the total for the variable-rate arrangement as described will be $473,730. This is quite a difference, to say the least, even though we were using very pessimistic assumptions. The amount is about what we would have paid for a fixed-rate mortgage with an interest rate between 15 and 16%.

Debt service payments for financing arrangements using such financing instruments as municipal bonds or the like can be calculated by using the same logic and formula if the bonds are organized like a mortgage. More often, the provisions call for periodic payments of a dividend over the period of the bond or loan and full payment of the original amount financed at the end (maturity) of the period. If the payments are to be generated from the revenues of the project itself (and not, for example, from the general tax revenues of the governmental entity), this means that a specific sum must be set aside each year in addition to the interest payments, so as to accumulate the full amount at the maturity date.

The extreme example above should make it clear that it is very necessary to keep close track of the anticipated pattern of debt service payments over the years. The owner is well advised to devote all possible effort to obtaining favorable financing terms for a building project; however, there is very little that the architect can do specifically, beyond making sure that the design is generally sound, attractive, and cost-effective, to help this effort.

Taxes

Several kinds of taxes must be considered as part of the cost of owning a building project. The main types discussed here are real estate taxes (property taxes), sales taxes on services, rentals, and so on, income tax on the income produced by the project, and capital gains taxes due on sale of the property.

Real Estate Taxes. These are local (city or county) property taxes based on the assessed value of the property—site and buildings. The determination of the value of a real estate project is made by a property appraiser. The tax is calculated by multiplying the tax rate by the assessed value. The tax rate is set by the local government and usually is expressed in "mills"—that is, pro mille, meaning dollars of annual tax for each $1,000 of assessed value.

The assessed value must be obtained from the property appraiser's office. The appraiser uses three basic approaches (or a combination of them) to determine the value of a building:

1. Replacement cost: this refers to what it would cost to replace the building with either
   (a) An equivalent one (one that serves the same purpose) or
   (b) An identical one (this is also called reproduction cost)
   Usually, approach (a) will be used; method (b) is appropriate for buildings of historical significance.
2. Market value (what similar buildings in the neighborhood sell for).
3. The present value of potential income the building can produce over its remaining useful life.

In many areas and municipalities, the assessed values of real estate differ from market values, in that properties systematically are assessed below reasonable market estimates. Real estate taxes and insurance often are grouped together as fixed expenses because they do not vary with occupancy and use.

Sales Taxes. Depending on the location of a property, there will be local and/or state sales taxes for rentals and lease payments. If the rentals are calculated on a cost basis (rent required to cover debt service, operation costs, developer profit, etc.), these taxes must be added to the rental rates.

Income Tax. Of interest here is the tax — federal and state tax where applicable — on the income (the cash throw-off or cash flow left after subtracting expenses and debt service from the effective gross income, minus depreciation allowance and replacement reserves) from the building every year. However, the rate at which income tax is calculated (the tax bracket) depends on the total income of a particular taxpayer and must be ascertained case by case. Also, tax rates may vary from year to year. For this reason, comparative assessments of the economic viability of building projects as it is influenced by design decisions are more meaningful based on "before tax" performance, whereas the owner obviously is concerned with the final "after tax" outcome.

Capital Gains Tax. Unlike the income tax, which is paid every year, the capital gains tax is paid only at the time of sale of the property, on the capital gain (i.e., the difference between the acquisition cost of a building and the sale price if the latter is greater than the acquisition cost). The tax law distinguishes between short-term (for property held less than a year) and long-term capital gains (for property held for a longer period). An exception is that anyone over age 55 selling a private residence is not liable for capital gains taxes for one such transaction during a lifetime. The capital gains tax rate also depends on the taxpayer's tax bracket.

Repair and Replacement Costs

Repair Costs. Building components not only require maintenance, but every once in a while will suffer breakdowns and need to be repaired. Experience allows us to estimate the average time between repairs for different types of components, and the average cost of the repair. The necessary repair funds usually are estimated as average annual costs that must be set aside in order to enable payment for the needed repairs when they occur.

Replacement Costs. If a building component cannot be repaired, it will have to be replaced. It may be difficult to draw a clear line between items often covered under maintenance such as the continuous replacement of worn-out or broken light bulbs, which is carried out by maintenance staff, and the more infrequent replacement of items such as broken windows, worn-out carpets, major appliances, HVAC systems, and building parts. Repair costs often are associated with at least some replacement and so are often listed with replacement costs.

Replacement Reserve. If the building must be replaced with a new facility at the end of its useful life, and the costs for the new building must be generated with the income from the old one, an annual replacement reserve must be provided for, to be paid into a fund set aside for this purpose. The rationale is the same as that behind the depreciation tax allowance: to ensure that sufficient funds will be available to provide for another facility, given that it may have to be replaced or sold at its reduced (depreciated) value. The amount that should be set aside is a matter of policy. At least, the amount for a down payment (equity contribution) for a new building of the same size should be considered. The tax laws provide rules for the annual depreciation allowance that can be deducted from the taxable income, but it must be distinguished conceptually from what an owner may wish to allocate to replacement reserves. Of course, the same principle can be applied to the replacement of major building components. Both types of replacement are estimated as annuities — equal annual amounts — paid into an account that is expected to accumulate to the required amount for replacement when needed.

In estimating replacement costs, both inflation and possible reinvestment interest rates must be considered.

As an example, consider the replacement of a building component: A roof is expected to last for about 20 years. It cost $100,000 when the building was built. Assuming an annual inflation rate of 5%, the single payment compound amount factor for 20 years is 2.6532 (computed with the formula or looked up in conversion factor tables; see Appendix 8), and thus the cost of replacing the roof after 20 years would be $265,320. If an annual amount set aside to accumulate to that sum were deposited in a savings account at 6% interest, the corresponding sinking fund factor would be 0.02719 (from tables), and the required annual amount would be $265,320 × 0.02719 = $7,214.05. If the savings interest rate were 5%, the sinking fund factor would be 0.03024 and the required annual amount $8,023.28.

Example for Replacement of the Entire Building: A building's initial cost was $1,000,000 at the time of construction. It is estimated to last for 30 years, at which time a new building must be built to replace it (accommodating essentially the same functions). At 5% annual inflation (single compound amount factor = 4.3218), it will cost $4,321,800 to replace the building at year 30. If it is then financed with a new mortgage at an 80% loan-to-value ratio ($3,457,440), an equity contribution of $864,360 will be needed. To find the annuity required to accumulate this amount in a savings account at 6% we must multiply $864,360 by the sinking fund factor 0.01265, obtaining $10,934. If we wanted to save up enough to pay for the entire replacement project outright, without taking out a new mortgage, we would have to set aside $0.01265 \times \$4,321,800 = \$54,670.77$ each year into the same 6% savings account—if we could not find a better reinvestment opportunity paying higher interest rates.

Renovation, Alteration, Addition Costs

To maintain a building's ability to serve its purpose over time, it is necessary to do more than just maintenance and repair/replacement of unserviceable components. As needs and purposes change, the building must be adapted to these changes. Renovations and alterations, redecoration, and, at the extreme, addition and expansion are different forms of such adaptation, and the costs of these measures must be included as regular elements of the cost of ownership.

Some of these costs should be anticipated to occur on a regular schedule, such as redecoration, painting, and the like, and could be included under the category of regular maintenance costs. More far-reaching alterations and remodeling will be less predictable, occurring in connection with major tenant changes or adjustments to shifts in market conditions. It is useful to include a regular fund for such expenses so that they can be paid for as they occur without causing serious disruptions of cash flow. Typically, the consideration of how such changes can be accommodated involves not only anticipating possible patterns of required changes, but trade-off decisions between building features that cost less but are less accommodating to changes and provisions that facilitate change but also tend to cost more (flexibility). The choice between fixed and movable partitions is the most common example. In an office building, for example, should the partitions between office cubicles be constructed of fixed studs and drywall (that would have to be demolished and entirely rebuilt in order to be placed in a different position) or made of movable, flexible panels that could be disassembled and reused in a different configuration? The latter usually cost more initially, but if the partitions have to be repositioned relatively frequently, the higher cost may pay for itself in a short time after only a few moves. This is a trade-off problem that can be analyzed by means of the decision theory approach, comparing the expectation value for the total present worth cost of initial construction and subsequent relocations as the performance measure. The example is discussed in Appendix 11, which deals with making decisions under risk and uncertainty.

Other examples of similar trade-off problems include the choice between different flooring materials (e.g., carpet versus vinyl flooring, where in addition to the cost of replacement for the purpose of remodeling, the cost of regular cleaning and maintenance for the two materials must be considered) or the issue of designing a building and its structural as well as other service systems to facilitate future expansion that may or may not happen. Providing for expansion in some way at first construction may add cost to the initial project that would be wasted if no addition ever were needed. Conversely, if the first phase were designed with no regard for possible growth, later addition could be considerably more expensive than providing for expansion initially.

Data for estimating costs of remodeling, alterations, and additions are available in various cost construction estimating manuals. The missing component for this task is forecasting the probabilities of whether the anticipated changes will be needed (e.g., the likelihood of having to move a given linear foot of partition in a given year). Past experience from other projects will be of only limited help in this, unless the project is a very routine one.

Miscellaneous Costs and Expenses

A number of other typical cost categories must be considered, ranging from vacancy and bad debt losses to contingency allowances.

Vacancy and Bad Debt Losses. These actually are reductions of rental income but usually are referred to as expenses or costs. Vacancy refers to the amount of rental space that is not occupied and rented at a given time, and which therefore does not produce rental income. Any change of tenant will cause a vacancy for some time, as the space must be cleaned, painted, refurbished, and so on. Expressed as an average percentage rate (of total amount of net leasable area, for example) or as its inverse, the occupancy rate (100% minus vacancy rate), it will reflect the overall attractiveness and competitiveness of the rental facility in the marketplace. Overall market conditions must be considered in estimating vacancy losses. For example, economic boom times have led to overbuilding in overly optimistic anticipation of increased demand in some areas. Thus, these regions have an oversupply of commercial office space—up to 35% or more in some instances. This means that a temporary overall vacancy rate of that much prevails in those regions and must be made the starting point of the projections. If the expected growth then does not materialize, the "temporary" high vacancy rate (which is expected and normal in any large project during the startup phase) will last longer and can contribute to the economic failure of an otherwise well-planned and sound project.

Traditionally, for a large multi-tenant office project, one might have assumed a vacancy rate of 50% or 40% during the first year or so, declining toward a steady-state rate of 5% or 10% over the next two or three years. These rules of thumb are no longer safe. To arrive at reliable estimates for a specific project, it is imperative to carefully study the actual economic conditions and the existing supply or oversupply of the planned type of facility in the project's neighborhood and region, as well as any general economic-climate projections, short-term and long-term, that might affect project performance. This is not necessarily the architect's responsibility, but the architect must be aware of these issues and ask the right questions of the client and the client's financial advisors to make sound assumptions in forecasting the economic impact of his or her own design decisions, which occur in the context of the overall economic conditions. Also, the design strategy with respect to the target share of the market might be influenced by these considerations; to become more competitive in a tight market overbuilt at the inexpensive end of the spectrum, the owner may ask the designer to shift the project toward a high-quality market, or vice versa.

"Bad debt" refers to rents not paid (or paid late) for premises that actually are occupied, or those owed by tenants who have moved out without paying the last few months' rent.

Contingency Allowances. Allowances for unforeseen expenses, damages, legal bills, and the like, should be included either as a per-square-foot figure or as a percentage of rental income.

Sale costs are the costs involved in selling property —either the realtor fees (usually 6% or 7% of the effective sale price) or the costs incurred in the owner's own efforts to sell the property: advertising costs and time spent in showing the building to prospective buyers, negotiations, and so on. Also, where applicable, they include sales taxes on the sale price of the building (which are different from the sales taxes on rents mentioned above).

Demolition costs involve the cost of tearing down the building if it must be removed, and of hauling away the debris.

NONMONETARY COSTS

The discussion thus far does not include the nonmonetary costs that may be associated with the planned project over its lifetime of use. Even if the economic impact of such costs is not directly visible, people often are quite willing to incur monetary costs in order to avoid or reduce such intangible costs—just as people are willing to pay for nonmonetary benefits. Or there may be indirect economic consequences of such intangibles later, so that it is a matter of common sense at least to identify these costs, perhaps making explicit what monetary cost differences would be involved in their avoidance or reduction and allowing those involved to make conscious trade-off decisions on their own.

Of course, raising this question opens up a wide field of concerns, ranging from architectural qualities and their relationship to user needs to questions of the wider environmental impact of a project: increase in traffic, loss of trees and vegetation, loss of views or sun incidence at certain times of the day for neighboring properties, pollution, stormwater runoff with the possibility of flooding in lower-lying areas downstream, and changes in neighborhood scale, character, and use patterns or property values, to name but a few such effects.

From the brief list of examples of the intangible costs of building projects, it also becomes clear that one of the problems is with the distribution of such costs over the set of parties involved in a project or affected by it in some way; although the monetary costs discussed above all are borne by the owner (or tenant) of the project itself (just as its benefits are assumed to accrue to the owner), many of the intangible costs are imposed upon neighbors or other members of society.

The question is: how should such costs be dealt with in the economic assessment of a building project? Appendix 6 discusses a number of different approaches to evaluation, some of which are particularly conducive to including noneconomic costs (and benefits) in the assessment. Benefit–cost analysis is one such method that translates noneconomic concerns into the common yardstick of monetary measure. Impact fees for new development are beginning to redirect the economic consequences of such societal impacts back to the developer or owner. It is likely that there will be more concern for the fair and equitable distribution of development impacts in the future, and legislation to ensure that unequitable effects can be corrected.

STRATEGIES FOR ESTIMATING AND CONTROLLING FUTURE COSTS-IN-USE

Estimating Costs-in-Use as an Integral Part of the Design Process

The task of estimating future cost of operating buildings, or the combined initial cost and long-term cost picture, will be called for at several points in the delivery process. Its results would be helpful any time there is a reason to recheck the feasibility and/or projected economic (long-term) performance in light of more detailed design decisions or other information. It should not merely be a perfunctory gesture after all decisions have been made, but should accompany the design process as an integral part. In doing this, some recurring questions must be addressed.

The first problem is that of finding out what cost amounts will occur at what points in time. This is in part a problem of prediction (see Appendix 10 on forecasting) of the costs in the future, taking account of the way the costs (as we know them from today's data) will vary over time—for example, as a result of inflation, or of other conditions in the market or in the nature of the cost to be predicted.

Comparing Costs at Different Points in Time: Comparing Solutions with Different Cost Distributions over Time

Another part of the problem is that of choosing a meaningful basis of comparing costs that occur at different points in time and of finding a single common measure for all these costs which can be used to compare several solutions that will have different patterns of cost distribution over time, and then to identify the least costly and thus the preferred solution.

There are several conventions for doing this. (The underlying issues of dealing with the time value of money are discussed in detail in Appendix 8.) The first is the present worth approach, in which all initial and future costs over the life cycle of the building are individually converted into their present value equivalents and then added up. The resulting measure is the present worth life cycle cost for a project. Another approach, the equivalent uniform annual cost or EUAC approach, is to convert all costs individually into their equivalent uniform annual cost and then add them up to a total uniform annual cost.

The resulting life cycle cost (LCC) of proposed alternative solutions can be compared in a straightforward manner, once the decision has been made as to which of the two methods to use, and the totals have been estimated in a consistent manner. The decision rule then would be to adopt the solution with the lower LCC.

However, the problem usually presents itself to the owner as a question of whether to invest in a more expensive system, as opposed to a conventional one (the base-line alternative). The owner then tends to see it as a question of whether the additional investment— the difference between the initial costs of system A and those of system B—is worthwhile. Its worth is measured in terms of the savings in long-term costs that will result from the more expensive system. This concern is addressed in the use of the savings–investment ratio (SIR), which is defined as the difference in the costs of the LCC for system A and that of system B over the difference in their initial costs:

$$\text{SIR} = (\text{LCC}_A - \text{LCC}_B)/(\text{ICST}_B - \text{ICST}_A) \quad \text{(5-1)}$$

In this case, the LCC should be expressed in terms of present worth. If the SIR for a proposed alternative is one or less than one, the solution is not worthwhile; if it is greater than one, the proposal is worth the extra investment.

Estimating Methods and Data for Costs-in-Use

As with initial cost, the methods for establishing estimates of future costs-in-use differ essentially according to the types of units to which projected unit prices are applied. The results are estimates that differ in their degree of specificity or detail. Again, the standard assumptions are that the greater the detail is, the smaller the range of possible error. For an estimate of long-term costs that is based on the area method, using the number of square feet of total floor area and prices expressed as dollars per square foot of floor area, it would be reasonable to expect an estimate within around a 15% error margin. A systems method estimate, broken down by the building's subsystems (but each of these expressed in terms of square feet of floor area to describe their magnitude), should be somewhat more precise; and a very detailed item-by-item estimate should reduce the error margin even more, all to the extent that the data used are reliable.

Data for cost-in-use estimates are not nearly so readily available and reliable as those for new construction, especially not in a form that would allow the architect easily to make meaningful design and trade-off decisions at the early design stages where the impact would be most significant. Besides a company's own record for older buildings, associations for specific building types such as shopping centers, hospitals, office buildings, and so on, compile such information. Public and government agencies keep records of this kind, most notably the U.S. Department of Defense, which has one of the world's largest (if not the largest) inventory of buildings under one ownership.

There is a special source of possible error in cost-in-use data. For example, the level of maintenance often is determined not so much on what should be done to keep the property in top condition, but on what is budgeted. And the maintenance budget could be the result of flawed estimates as well as deliberate policy decisions to keep maintenance costs down (and profits up). This leads to a vicious cycle because the trimmed-down maintenance accounts end up as data that will be used to make maintenance estimates for future projects, and so on.

The effects of such self-reinforcing mechanisms can be seen, for example, in some European countries where visitors may be puzzled by the contrast between the unfinished, apparently unkempt exterior of many apartment buildings and their quite modern, well-appointed, and tasteful interiors. The explanation is that property taxes are assessed on the basis of how a building looks from the outside. This gives property owners a powerful motivation to defer external upkeep. The costs for such upkeep then do not need to be charged to tenants; and because this pattern is so pervasive, the resulting low maintenance figures become the data basis for future feasibility estimates, which in turn determine the operating and maintenance budgets—which then predictably do not contain more than minimal upkeep funds.

Controlling Costs-in-Use

What means are available to the architect for influencing the long-term ownership costs of buildings? The first vital step is simply that of becoming more aware of these costs, and of their relationships to design decisions. The second step, no less important and already stressed above, is actually to perform repeated estimates and analyses of such costs for competing design solution alternatives throughout the planning process —that is, to make the Life-Cycle Cost Analysis, or the examination of annual expense, an integral part of the programming and design process, not just an added token gesture that does not inform design decisions. (Life Cycle Cost Analysis is discussed in more detail in Chapter 8.) For this, it is necessary that the analysis be adapted to the various stages of the process and to the level of detail and reliability of data available at each stage—much like the estimating methods for initial cost. With computers becoming more common tools in design offices, and CAD and other software becoming more integrated, it is foreseeable that viable life cycle cost analyses can be performed automatically even for schematic design proposals. It then will become much easier to quickly try out different design solutions, materials, details, and so on, and to equally quickly obtain the initial as well as long-term cost implications for each alternative, selecting the best decision at each stage and building it into the more detailed solution in the next phase. In the meantime, being able to apply currently available software and data—for example, in constructing and using simple ad hoc spreadsheet models for the analysis of alternative design solutions, especially at early stages—will go a long way in controlling these costs.

ESTIMATING COSTS-IN-USE: CONCLUSIONS

In this summary of some of the difficulties that arise in making cost-in-use estimates over the long term, it should be noted that many of these conclusions draw on issues discussed elsewhere in the book. Problems facing the architect who tries to deal with costs-in-use as design considerations in a responsible manner are as follows:

Conceptual Distinctions

It is difficult to distinguish and separate the various elements of costs-in-use in a clear, unambiguous, and consistent manner. There are several areas of uncertainty as to how cost positions should be classified and grouped together. (For example, utility bills contain several items that have to do with energy use; or there are conceptual overlaps between maintenance and repair costs, especially if some of the work involved is done by personnel combining managerial and maintenance/replacement responsibilities.) This uncertainty makes it difficult to interpret historical data adequately.

Data Problems

Another main concern is that of data availability: useful data often are not available at all, or are not in the format needed to support design decision-making. For example, there is a shortage of useful data that might facilitate meaningful trade-off decisions at the early schematics phase. If data are available, it often is not at all clear what the figures mean. They must be seen in the context of inflation trends and cost-of-living developments, and interpreted with regard to changing standards of quality of materials and finishes, appliances, level and quality of service, and patterns of use. The costs-in-use records for a house built in the 1960s will be quite different from those of a house completed in the early 1990s. All the problems of inflation, price escalation, and seasonal fluctuations, as well as those involving the compilation, scrutiny, analysis, and interpretation of statistical material of any kind, come into play here.

Measuring Performance with Respect to Cost-in-Use

Because of the two problems mentioned above, it is not easy to develop clear-cut, reliable, and consistent measures of building performance with respect to costs-in-use.

Forecasting Costs-in-Use

Because costs-in-use considerations for building plans are concerned with the future, all the difficulties discussed regarding the interpretation of past data are mirrored in the task of forecasting or predicting what the corresponding trends will be in the future. As indicated in more detail in Appendix 10, it is difficult enough to forecast phenomena that are based purely on laws of nature, without human interference. In contrast, almost all variables under scrutiny for future costs-in-use are less governed by immutable and consistent natural laws than driven by human intentions. Attempts in economics, political science, and other disciplines to develop models for forecasting trends in human affairs always have run straight into the problem that people interfere with trends by reacting to forecasts as much as to the trends themselves. And they do so sometimes by following the trend and therefore reinforcing it, affirming the forecast (the proverbial "self-fulfilling prophesy"), or sometimes by going against the forecast and thereby refuting it. Here people face decision situations characterized by risk or uncertainty that sometimes are rooted in the nature of things and sometimes are caused by the intentions, plans, and decisions of other people—who may themselves try to anticipate what the others are going to do. Dealing with such situations, people reveal themselves to be of quite variable psychological makeup with respect to their degree of optimism or pessimism and their willingness to take risks or to devote extra effort to gathering additional information and critically examining the information. The eventual decisions depend as much on these factors as on the objective reality in which they operate.

These difficulties have led many scientists (such as Herbert Simon) and economists to despair of human rationality as the underlying guarantee of their forecasting models, introducing notions such as "bounded rationality" or even inherent reliance on intuition (that is, nonrationality or even irrationality) into their theories as explanations for the failure of forecasts to materialize in reality. It would perhaps be better to start from the premise that human beings by nature are trying to change the reality they find around them, that they do so by developing a number of competing designs—not obeying some immutable predetermined law—and then adopting one of them. In a group or society, several conflicting designs often are pursued simultaneously. This premise does not force us to deny human rationality, but it adequately explains the difficulties in predicting trends in human affairs in any systematic and "objective" fashion.

The influence of policy on future outcomes is shown clearly, for example, in decisions relating to trade-offs between initial and long-term costs, or in the problem of allocating funds to maintenance, upkeep, and repair programs. These decisions are not only a response to inevitable deterioration processes but invariably are already expressions of attitudes about a building and its life, about the mutual adaptation of the building to its users and vice versa, and about time, preferences, and priorities.

Connections between Cost-in-Use Measures and Design Decisions

A further difficulty lies in the lack of clear-cut, well-defined, and usable knowledge about the relationships between cost-in-use measures and architectural design decisions. This is not to say that such connections do not exist—they definitely do—but that their effect is difficult to pinpoint with the desired accuracy, both qualitatively and quantitatively. This difficulty is in part due to the multitude of variables that interact in many ways, so that changes in outcome always can be explained by several factors; in part due to the considerations discussed above; and in part due to the way data are being provided, which does not always differentiate them clearly enough for the design decisions to be made. An example, again, is the lack of data relating architectural form decisions to effects on both initial cost and costs-in-use.

These few observations might lead one to the conclusion that reliable long-term forecasts of costs-in-use are on such shaky ground that they are not worth the effort. This would be an inappropriate conclusion. The fact remains that architectural design decisions do have a significant impact on long-term economic costs of buildings, and that the architect therefore must bear some responsibility to the client for these consequences. But the way in which this responsibility is dealt with will be different from, say, an engineer's forecasts regarding the structural stability of the building, which is determined almost exclusively by "objectively" determinable natural phenomena, which can be predicted and modeled by the expert (structural engineer), who then can assume responsibility for the resulting decisions. The client's opinion does not enter into the engineer's calculations. This is not true of the decisions influencing long-term costs; here, the architect must engage in a discourse with the client (and users), in which both the potential impact of the design decisions and the role of the client's own assumptions, attitudes, policies, and expectations as they operate within the spectrum of possible changes in the context, which neither the architect nor the client expects to be able to alter.

STUDY QUESTIONS

1. For a loan of $120,000 compare the following arrangements:
   (a) Fixed-rate, 30-year conventional mortgage at 10% interest and 2% discount points (the loan fee or discount points are an added up-front cost).
   (b) Fixed-rate conventional 20-year mortgage at 10%, with 3% discount points (fee).
   (c) Same as (b) but with 9% interest and 4% points.
   (d) Variable-rate, 20-year, 1-year-adjustable loan at 8% interest with an annual cap of 1% (i.e., the interest rate cannot rise or fall by more than 1% per year, with an upper limit of 18% and a lower limit of 6%).

   For each, find the annual or monthly payments as well as another meaningful basis for comparison.

(e) Assume that an owner can afford to pay $800 in monthly mortgage payments. How much money can she or he afford to borrow at 9% interest? Assuming further that the terms of the lending institution include a loan-to-value ratio of 0.8, what is the maximum cost of the project? What if the owner had no more than $15,000 to put down as equity contribution for the building?

2. Copy the spreadsheet shown in Figure 5-4 (A and B). Develop a hypothesis as to what design decisions you might modify to bring the costs-in-use down by 10%. Then test the hypothesis by varying the corresponding numbers in the spreadsheet. Compare your results with the changes in context variables (mainly financing terms) that would achieve the same overall result.

6
Benefits and Value of Buildings

OVERVIEW

The discussion in the preceding chapters was about cost; it adequately reflects the reality of concerns that will dominate the work in actual projects. However, focusing exclusively on costs in building economic performance is only part of the picture, and may even be counterproductive. The point is to relate the costs to the values or benefits received in return for the costs; the concept of economic performance tries to express this more general concern. This chapter will look at some of the monetary as well as nonmonetary benefits that are typically obtained from buildings, and at how to estimate and control these benefits.

There are several basic distinct forms in which economic rewards can be obtained from building projects; these are recognized in real estate theory and law as the different forms of tenancy. The first is the economic value of using the building for the owner/investor's own operations or enjoyment, as in a home, or a company's own headquarters. The benefit is entirely intangible in the case of a dwelling, and indirectly economic in the case of space used for the production of goods or services. The legal relationship of the user to the property in this case is that of freehold. Second, the building can be sold to somebody else; the freehold tenancy is transferred to a new owner in return for a sale price. Here, the economic benefit is entirely monetary and focuses on the profit or excess of money received in the sales price over the cost the seller originally invested in the project. The third possibility is that of leasing or renting the whole or parts of the building to others, in return for periodic (daily, as in a hotel, monthly, or annual) payments. Although owner retains the freehold

relation to the property in this case, the tenant or actual user is in a leasehold tenancy. There also are combinations of relationships that can be specified in contracts —for example, the lease with an option to buy, where the relationship on the part of the user begins as a leasehold, but can be changed into sale and freehold at specified points in time, in which case the lease payments already made are counted toward the sale price.

The monetary benefits derived from a building thus are as follows: direct money payments in the form of the rent or lease or the sale price; indirect economic benefits derived from the production of other amenities on the premises; and the economic equivalent of the intangible value of occupying and enjoying a building. Here, we are primarily concerned with the question of how the different forms of revenues or benefits can be predicted and estimated, how they relate to initial and long-term costs, and how these estimates relate to architectural design decisions.

MONETARY BENEFITS OF BUILDINGS

Sale Price and Profit

Many investors in building projects, and many developers, are in business in order to sell the buildings they build as soon as possible after completion. They make their living from the profit on these buildings, that is, the difference between the cost they had to invest in getting the buildings built and the sale price the buyers are willing to pay. How can one anticipate what that price will be? The problem of answering this question is discussed in the literature as the problem of the appraisal of property value, both for the purpose of estimating value for buyers and sellers in the real estate market and for the purpose of establishing property value as the basis of real estate taxes. Three basic approaches are recognized: the cost approach, in essence equating value with the cost of producing a building project; the market approach, which looks at what buyers of similar projects in the same area have been paying; and the income approach or income capitalization approach, which focuses on the economic value of the stream of income a project is expected to generate over time.

A simple answer to the value appraisal question would be obtained by starting with the known acquisition or construction cost, adding a profit at some specific rate, say, 12%, 15%, or 20%, to the developer's cost, and letting the total be the sales price. But profit added to what? Should the profit be based on the total cost of construction of, for example, $100,000 for a family dwelling? In this case, the profit at a 12% rate would be $12,000, and the total price to the buyer $112,000. Or should it rather be based on the developer's actual equity investment in the project? This would be about $20,000 if the lender used a loan-to-value ratio of 80% for the construction loan. (Although many projects are financed at 100% or close to 100%, the developer's efforts in researching the market for the project, assembling the financing package, and so on, represent a considerable and risky up-front investment, which is difficult to quantify.) In this case, the profit would be 12% of $20,000 = $2,400, and the total price $102,400. Or should it be a profit on the equity plus a charge for the time and effort the developer had to devote to the project (which sometimes are actually itemized as "development costs" in the project cost)? One can find all these attitudes, and everything in between, in the real estate market today.

However, what owners or developers would like to charge is mitigated by competition and available buying power in the market. And even if a developer eventually gets the asking price for a building based on one of the above formulae, if he or she has to wait many months for a sale to go through, then the value of that price certainly is reduced. In deciding what they are willing to pay, buyers do not really care what the developer's profit is but are concerned with whether they are getting their money's worth. So what determines the buyer's perception of what is a fair or good price? The economic assumption at the outset is that the original cost is an adequate but approximate measure of the value of the project; the actual value to the buyer must be higher and at least include the seller's markup over the original cost. So, many assessments of the question of economic benefit of building profits start with this assumption. But then, all the factors of the market come into play to complicate the estimate. Inflation and appreciation are examples of such factors, and for simplicity the two often are considered together. Then one can estimate the sale price at some future year n ($SALPRC_n$) by applying the appropriate factor for inflation (and appreciation) (INFL) to the original known sale price (or cost) at year 0:

$$SALPRC_n = SALPRC_o * (1 + INFL)^n \qquad (6\text{-}1)$$

if sale prices as known today ($SALPRC_o$) are expected to rise annually with inflation. For longer periods, building obsolescence and depreciation must be taken into account, and the site value still may appreciate even faster than general inflation. A formula that expresses this is:

$$SALPRC_n = BLDVAL_o * ((s - n)/s) \\ + SVAL_o * (1 + SAPPR)^n \qquad (6\text{-}2)$$

where:

$BLDVAL_o$ = The value of the building in year 0 ("now")

$SVAL_o$ = The value of the site in year 0

$SAPPR$ = The rate of appreciation of the site (which may or may not be different from general inflation)

s = The expected life span of the building

The value of the building is closely related to the level of maintenance of the building every year. Aside from obsolescence due to different user standards, fashions, and technological advances, it is possible to keep a building in full income-earning capacity indefinitely; the annual maintenance amount needed for this is called the 100% maintenance level. Usually, buildings are not maintained at this level, and, as a result, at some point the required level of annual upkeep will exceed the income—especially if rental income does not increase as fast as maintenance costs rise.

Another factor that will determine the future sale value is supply. How many similar buildings are available in the area? What is the average occupancy rate for buildings of this type in the area at the time of the sale? How desirable is the area going to be? How much would one have to pay to buy or rent a similar building? What are the costs of operating and owning the property in the long run? What are the overall projections regarding the rise or decline of property values in the area?

One of the most significant aspects in real estate valuation is location. The selling price for the same size and quality house within the same city may vary by as much as 100% or more, depending on the neighborhood in which it is located. Many factors are responsible for this, only some of which can be approached by straightforward rational means—for example, the distance from downtown and travel costs. If a family can afford to pay $800 per month for housing, and a specific location would require spending an extra $150 per month on transportation, the available monthly amount the family can spend on housing in that location would be $650. Other considerations are: the quality (perceived or real) of schools, shopping, and recreational amenities; the proximity and/or visual impact of natural features such as parks, bodies of water, or vistas, or of undesirable areas such as industrial areas and so on; the age of a neighborhood (how new or well established it is; the style, mix, and upkeep of the buildings; the landscaping); age, economic, and ethnic demographics; the crime rate; and outright fashion and social status considerations. All combine to influence the selling price of real estate property.

Finally, for commercial properties, there is the income capitalization approach: How much money can one expect in rental income from the building? The capitalized value of that income stream—either the present value of the expected income over a specified number of years, at a specified discount rate, or the annual income multiplied by the net income multiplier—must be considered the most a buyer should pay for a project. This brings us to the next question, that of the rental income one can expect from a building.

Income-Producing Properties: Lease and Rent

Most of the market-related questions that make prediction of sale price so difficult, also pertain to the anticipation of income from leasing or renting buildings. Again, reliable predictions about "what the market will bear" can be obtained only from thorough analysis and knowledge of the market conditions prevailing in the area where the building is (to be) located. We can calculate the level of rent that must be charged in order to recover the costs of construction, financing, and so on, to "break even." (This is the "front door" approach to financial feasibility analysis, to be discussed in detail in Chapter 8.) Beyond that, things become uncertain. This uncertainty is manifested in phenomena like the vacancy rate (that percentage of the rental space in a building that is not rented and producing income at any given time because of tenant turnover and delays in getting a new tenant, or bad checks, late rent payments, etc.). If an owner sets rental rates too high, he or she may get some tenants eventually but will pay for doing so with a high vacancy rate. Comparison with prevailing rates on the market is vital to arrive at realistic rental projections.

In estimating income from leases and rentals from a building, the proper interpretation of market information is essential—for example, what is included in the services provided for a lease (see Appendix 13 for a summary of different lease types.)

Regardless of the various provisions made regarding the lease, it is always important to include the results in a larger picture visualizing the overall relationships of benefits and income in time to the costs. A cash flow diagram summarizes this (Figure 6-1).

Technically, assuming that we know the appropriate values of sales prices, equivalent rentals, and the pertinent inflation, appreciation, and depreciation factors, the future values of rentals and sales prices can be easily estimated by using the appropriate time conversion factors.

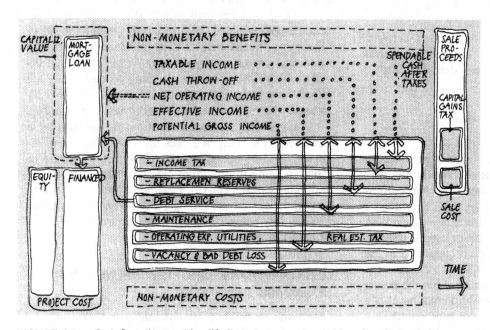

FIGURE 6-1. Cash flow diagram (simplified)—monetary costs and benefits of rental projects.

Calculation of Rental Income

Rents are paid monthly, but for the purpose of estimating economic worthwhileness are calculated as annual payments. Given the established rent for the annual rent "now," year 0, the rental for a future year n will be:

$$RENT_n = RENT_o * (1 + INFL)^n \qquad (6\text{-}3)$$

if rentals are raised with inflation every year; otherwise appropriate provisions must be made. For example, leases are often agreed upon at fixed levels for a period of, for example, three years, five years, or even more. If the rentals are "triple net," that is, covering only the owner's mortgage payment, the rent could in theory be fixed for even longer, up to the mortgage term. Note that this is gross rental income before applying vacancy and bad debt losses. This is considered by many to be an expense (cost); the net rental income or net effective income would be:

$$NETRENT_n = RENT_n * (1 - VR_n) \qquad (6\text{-}4)$$

where VR_n is the vacancy rate for year n.

In larger multi-tenant projects, especially in new hotels and the like, it sometimes takes a considerable amount of time before the building reaches the expected stable occupancy level. The income projections therefore should consider lower occupancy rates during the "lease-up" period; for example, 60% (corresponding to a vacancy rate of 40%) during the first year, 80% during the second year, and approaching the stable state occupancy of 90% or 95% in the third year after completion. For hotels and similar building types, there will be significant variations in the rental income for different seasons. The rate structure for vacation hotels, for example, recognizes this by offering different rates during peak season, off-season, and the in-between "shoulder season"—an attempt to trade occupancy (volume) against rental rates to keep demand steady.

NONMONETARY BENEFITS; USE VALUE OF BUILDINGS

When an owner occupies and uses a building either for residential purposes or to house the operations of a business, there is no cash transaction involved in doing so. The enjoyment of the building, even the image or prestige benefit to corporations from new impressive corporate headquarters, is to a large extent an intangible. How then can the value of such occupancy and use be determined? The economist's answer to this is somewhat tautological: its value is precisely what the owner would be willing to pay for the use and enjoyment of a building with the same features on the rental market—which is presumably equivalent to what he or she was

willing to pay in terms of present worth equity investment and ongoing mortgage payments (plus operational expenses) when the decision was made to build in the first place. The answer seems to suggest, again, that value is equal to cost; but this is no help at all, as we precisely want to find out what cost is acceptable in return for what value. If the definition of value was indeed cost, we could not distinguish between alternative design solutions. It also does not explain the discrepancies we can observe in the real estate market between the cost developers incur to build buildings and the prices they charge to sell them. So value must be defined—and measured—by something other than cost on one hand, and sale price or rental income on the other. This question raises all the problems of evaluation that are discussed in Appendix 6, and there is no easy way to resolve them. The standard techniques of feasibility analysis all but ignore the problem by substituting area measures (square footage) and at best call for application of judgment based on experience in the setting of rental rate assumptions, which then are multiplied by the square footages.

Persistent questions about the basis for such judgment will reveal that there are indeed many design features that account for differences in valuation. Real estate listings refer to such items (things) that may be present in a building: fireplace, number of baths, number of bedrooms, built-in closets, garage, swimming pool, to name but a few. For office buildings sometimes one finds mention of corner offices as a desirable feature. Views and of course location always figure prominently.

Overall though, the discussion of the relationship between value, costs, and income may leave many with a sense of dissatisfaction; it is difficult to see the connections, if any, between these variables and architectural design decisions that presumably create the value. One can establish lists of desirable (and undesirable) building features and perhaps correlate them statistically with prices people are willing to pay. Research efforts have used game situations with trade-off decisions to find out what people really value in housing, for example. But the precise relationship between these values and the rents or prices people will pay remains elusive. A different approach to capturing a sense of building value that is more closely related to the kind and quality of the activities that are accommodated by a building, as well as its image, is discussed in Appendix 12 (building value measures based on quality of occasion and image).

DESIGN STRATEGIES FOR INFLUENCING ECONOMIC BENEFITS

Given the somewhat unsatisfactory outcome of the above attempts to pinpoint definitions and measures of noneconomic benefits from buildings, which in the end also determine the economic ones, a survey of strategies that can be recommended for the control of economic benefits through design measures is not likely to be much more conclusive. As long as the most common measurement used for estimating rental returns and sale prices is the floor area (net leasable area), guidelines for design will be largely limited to rules of thumb for maximizing leasable floor area or maximizing the net-to-gross ratio—at least as long as the strategy is looking at the first level of income measures: gross potential income, or even effective income.

Things get a bit more interesting when the focus shifts to Net Operating Income (see Figure 6-1) or Cash Throw-Off, as these measures are dependent on costs (both initial costs through debt service) and costs-in-use, and especially the latter can be manipulated by a number of design variables. Some of those issues were discussed in Chapter 5.

Common sense tells us that the real impact of design on benefits or building value lies in paying attention to those many details that have to do with how the building meets functional, emotional, and practical but also aesthetic user needs, which make a building enjoyable

—with the technical adequacy required to meet environmental challenges such as climate and with structural safety, ease and convenience of maintenance, and so on; in short, all the things that make good architecture. The problem is that all those design features are difficult to connect with economic value in a direct quantitative fashion that can be expressed in a mathematical formula such as the above equation describing the impact of inflation of future sales price. This leads to an understandable but problematic tendency to give more weight to those factors that can be dealt with in a quantitative manner than to others, with the consequence that the impression sometimes arises that concern for economic issues is equivalent to ignoring or undervaluing other quality aspects—that an economical building equals ugliness. Or that the only reason economy-minded corporations would be interested in architectural design would be the advertising value of recognized name designers, stressing the visibility and fashion aspect of such corporate image architecture rather than its inherent quality and value to the user. Nothing could be further from the truth; properly understood, architectural value is also economic value. However, it seems to be difficult to communicate this

EXAMPLE - INITIAL AND ANNUAL COSTS-IN-USE; RENTAL REVENUES

ASSUMPTIONS

Program/given			unit	formula
Net leasable area	NLA	80,000.00	sf	
Site area	SA	200,000.00	sf	
Assumptions				
Corridor width	CW	6.00	f	
Exterior wall thickn.	WTH	0.50	f	
Floor-to-floor height	FH	10.00	f	
Office depth	OFD	20.00	f	
Staircase width	STW	10.00	f	
Elevator lobby width	ELOBW	10.00	f	
Parking requirement std.	PARQ	200.00	sf NLA / prk sp.	
Parking spaces needed	PARSP	400.00	NLA/PARQ	
Parking area/space	PARSPQ	280.00	sf/space incl. drive	
Parking/drive area needed	PARKAR	112,000.00	sf	PARSP*PARSPQ
Landscaping requirement	LSRQ	0.10	% of SA	
Landscaped area required	LSARQ	20,000.00	sf	LSRQ*SA
Number of floors	NOF	5.00	#	
Leasable area / floor	LAF	16,000.00	sf	NLA/NOF
Office width (total)	OFW	400.00	f	LAF/(2*OFD)
Building length	BL	421.00	f	OFW+STW+2*WTH+ELOBW
Building width	BW	47.00	f	2*(OFD+WTH)+CW
Building footprint	BFP	19,787.00	sf	BL*BW
Total Floor Area	TFA	98,935.00	sf	BFP*NOF
Actual landscaped area	ALSAR	68,213.00	sf	SA-BFP-PARKAR
Exterior wall area	EWALLAR	46800.00	sf	NOF*FH*2*(BL+BW)
Glazed/opaque ratio	GOR	0.40	%of EWALLAR	
Glazed area (windows)	GLAR	18720.00	sf	EWALLAR*GOR
Opaque wall area	OQAR	28080.00	sf	EWALLAR-GLAR
Office module width	OFMW	20.00	f	
Number of off. partitions	OFPART	20.00	f	OFW/OFMW
Interior wall area	IWALLAR	24210.00	sf	NOF*((OFPART*FH*OFD)+(2*BL))
No. elevators per flr.	NEL	2.00	#	
No. elev. landgs	NELNDS	10.00	#	NEL*NOF

Data assumptions

Variable	Symbol	Value	unit
A/E Fee percentage	AEFEEPC	10%	%of CCST
Development Price	DEVPRC	1	$/sf of SA
Legal &Acc. fee perc.	LEGFEEPC	3%	% of CCST
Cnstr. contingency %	CCONTPC	0.05	%of CCST
Permit percentage	PRMPC	0.02	%of CCST
Real est tax mills	MILLS	0.025	% of SCST
Construction period	CP	15	months
Constr.loan-value rati	CLVR	80%	of SUBTOT
C. loan average			
outst. balance percent	AVOBPC	60%	of CLOAN
Constr. loan interest	CLINT	12%	of CLOAN/yr
Discount points	DSCPTSC	0.03	%ofCLOAN
	DSCPTSM	0.02	%ofMLOAN
Mortg. interest rt	MLINT	0.095	%of MLOAN/yr
Mortgage term	M	25	years
Mortg. ln/val rat.	MLVR	85%	of TIPRJCST
Equity discount rate	DISC	0.12	%annual
Rental rate	**RENTR**	**$25.00**	**$/sf/yr**
Vacancy rate	**VR**	**9.00%**	

FIGURE 6-2A.   Rental income projections for office building. Input assumptions.

56 Benefits and Value of Buildings

FIGURE 6-2B. Rental income projections for office building. Initial and annual cost and income.

Area / element	Symbol	Value	Unit	Initial price Symbols	Value	Initial cost (...CST)	Formula	Operation	Mainten	Repair & Replacemnt	Total annual
Site area	SA	200,000.00	sf	SPRC	$5.00	$1,000,000.00	SPRC*SA	$0.05	$0.02		$0.07
Landscaped area	ALSAR	68,213.00	sf	LSPRC	$5.00	$341,065.00	ALSAR*LSPRC	$0.02	$0.50	$0.20	$0.72
Parking &driveways	PARKAR	112,000.00	sf	PARPRC	$4.50	$504,000.00	PARPRC*PARKAR	$0.10	$0.35	$0.30	$0.75
Serviced space (floor area)	SSP	98,935.00	sf	SSPRC	$37.50	$3,710,062.50	SSPRC*SSP	$2.00	$0.80	$1.00	$3.80
Foundation/ground contact	FNDAR	19,787.00	sf	FNDPRC	$5.00	$98,935.00	FNDAR*FNDPRC	$0.00	$0.01	$0.10	$0.11
Roof area	RFAR	19,787.00	sf	RFPRC	$7.00	$138,509.00	RFAR*RFPRC	$0.30	$0.03	$0.10	$0.43
Glazed area (windows)	GLAR	18,720.00	sf	GLPRC	$21.00	$393,120.00	GLAR*GLPRC	$2.00	$0.20	$0.15	$2.35
Opaque extr. wall area	OQAR	28,080.00	sf	OQPRC	$16.00	$449,280.00	OQAR*OQPRC	$0.75	$0.10	$0.15	$1.00
Interior wall area	IWALLAR	24,210.00	sf	IWPRC	$2.50	$60,525.00	IWALLAR*IWPRC	$0.00	$0.10	$0.05	$0.15
Elevators	NEL	2	#	ELPRC	42000	$84,000.00	NEL*ELPRC	$200.00	$600.00	$1,500.00	$2,300.00
Additional landings	ADDLND	6	#	ALNDPRC	$7,000.00	$42,000.00	ALNDPRC*ADDLN	$50.00	$60.00	$100.00	$210.00
							Annual expense & oper. ANEXP				$930,520.30
Subtotal building	BLDCST					$4,976,431.50	Annual/sf TFA				$9.41
Construction cost	CCST					$5,821,496.50					
Constr. contingency	CCONT					$291,074.83					
Subtotal	SUBTOT					$7,112,571.33	SCST+CCST+CCONT				
A/E Fees	AEFEES					$711,257.13	SUBTOT*AEFEEPC				
Permit costs	PRMCST					$116,429.93	PRMPC*CCST				
Real est.taxes	PROPTX					$25,000.00	MILLS*SCST				
Carrying charges						$1,000.00					
Project Cost	PRJCST					$7,966,258.39					
Constr.ln interest amnt	CLINTAMNT				$512,105.14		SUBTOT*CLVR*AVOBPC*CP/12*CLINT				
Constr. disc.points	CLOANFEE				$170,701.71		SUBTOT*CLVR*DSCPTSC				
Constr. financing cost	CFINCST					$682,806.85					
Tot. Init. Proj. Cost	TIPRJCST					$8,649,065.23					
Mortgage Loan	MLOAN				$7,351,705.45		TIPRJCST*MLVR		Debt service DS		$778,982.24
Cash equity contribution	CEQC				$1,297,359.79		TIPRJCST-MLOAN		Total Annual cost until yr 30 w/o equity		$1,709,502.54
									Equity Uniform Annual Equivalent		$165,413.33
RENTAL INCOME									Total annual incl. equity		$1,874,915.88
Gross pot. rental income	POTREV		$/yr.	$2,000,000.00			NLA*RENTR				
Vacancy loss	VAC		$/yr	$180,000.00			VR*POTREV				
Effective income	EFFINC		$/yr	$1,820,000.00			POTREV-VAC				
Net Operating Income	NOI		$/yr	$889,479.70			EFFINC-ANEXP				
Cash Throw-off	CTO		$/yr	$110,497.46			NOI-DS				

STUDY QUESTIONS

and to get this message across when it must compete with mathematical formulae and spreadsheet calculations. Much work remains to be done on this issue.

1. A recurring theme in the above discussion was that economic monetary benefits ultimately depend on nonmonetary benefit assessments. Discuss this.
2. Figure 6-2 presents a rental income projection spreadsheet for the same project as the one shown in Figure 5-4 for costs-in-use. Identify selected design variables that you think might improve the rental income from this project, and manipulate them to see how much they do influence revenues.
3. Think of ways in which you can introduce distinc-

tions among types of leasable floor space in a building, for which different levels of rental rates might be charged (natural lighting? flexibility? dimensions?). Does this give you ideas for design changes you might consider to improve the overall attractiveness and income-producing capability of the building? Try some of these changes and check their impact by modifying the spreadsheet accordingly.
4. Test the influence of location on the income projection of the project by assuming a different site of the same size and twice the cost as that shown in the spreadsheet. How much more rent will you have to charge to offset the resulting higher project cost? What if the change in location also justified a more favorable vacancy rate—say, 8% or 5% rather than 10%? Discuss your answer.

7

Relating Building Costs to Benefits: Measures of Economic Performance

SURVEY OF ECONOMIC PERFORMANCE MEASURES FOR BUILDING PROJECTS

The preceding discussion of costs of buildings (Chapters 2, 3, and 5) and benefits and value derived from buildings (Chapter 6) may have taxed the patience of those who sensed the shortcomings of costs and/or benefits by themselves as measures of the worthwhileness of building projects. It is time to look for more meaningful measures of economic performance of buildings, specifically, measures that relate costs and benefits to one another. There are a number of different kinds of such measures. This chapter offers a survey of such measures of economic performance of real estate investment projects, discusses their relationship to owner investment objectives, and how architectural design decisions can influence these measures.

The discussion assumes an understanding of the concepts of measurement and variables and an understanding of the systems view of design; specifically, of the three types of variables by means of which design problems can be described: design, context, and performance variables. The reader also should be familiar with the concept and mechanics of time value of money: converting present to future, future to present value, and so on, as described in detail in Appendix 8.

The problem of assessing the worthwhileness of building projects, from an economic point of view, is equivalent to that of selecting an appropriate measure of economic performance and then estimating its value for the project under scrutiny. A measure of performance was defined as a variable that can be measured, that tells the designer of any object or solution for a problem how well a solution is doing, how well the object or solution serves its purpose or solves the problem.

A measure of economic performance for a building is a variable that measures how well the building meets the economic objectives and concerns of the client, owner, or user. To some, the use of the term "performance measure" for such variables as cost may seem confusing because performance often is understood to be the level of function or benefit one receives in return for the cost. However, if "doing well" (to a client) means keeping cost down, then cost is a performance variable. If it means having a large amount of money from rental income left over after paying mortgage payments and current expenses, then a performance measure must be used that is made up of those variables.

Types of Economic Performance Measures for Buildings

There is no single measure of economic performance that can be used for all buildings in all situations. We distinguish a number of different types or groups of such measures that have been proposed and used. They range from simple, direct dollar-amount-based measures to complex ratios involving several variables, rates of growth, and breakeven periods. Within each group, there are several individual measures.

The main types of economic measures of building performance are roughly the following:

- *Costs* (initial or first costs, continuing or running annual cost, life cycle cost, etc.): Costs are almost always understood as measured in money, and our discussion thus far has been restricted to this understanding. However, there are some other approaches that can be considered variations of the cost criterion but which use different units of measurement—for example, measuring the cost of a building (both initial and especially operating costs) in units of energy (e.g., Btu's) or hours of labor. The energy crisis of the 1970s increased people's concerns with the consumption of energy needed for operation but also for the construction of buildings. Making up energy budgets for buildings can yield interesting, even dramatic, insights into many assumptions we usually take for granted but which sometimes turn out to be quite questionable. Studies of costs of living in different countries often resort to comparing the number of hours the average wage-earner must work for a given item, for example, housing. These measures also encounter the problem of converting units from one currency into the other—how energy and material costs are translated into labor, or vice versa, for example. The traditional approach of translating all such measures into money equivalents (so as to arrive at common measurement units, e.g., dollars) can accommodate the concerns behind these approaches even if the resulting budgets tend to hide rather than reveal the energy waste or the inequity in labor wages inherent in a given project. Of course, the implied argument behind poposals to use energy measures, for example, is that current energy prices are unrealistically low and will rise dramatically in the future.

- *Benefits:* Included here are revenues, returns, profit, income, and cash flow, as well as reversion (selling price); but also use value, and aesthetic and other intangible benefits. Savings in monetary or nonmonetary costs, between standard baseline solutions and proposed alternatives, also are often treated as benefits.
- *Ratios of benefits or revenues over costs, and rates of investment growth* (the two are conceptually distinct but closely related).
- *Periods of time spans within which a given objective (as measured by some other criterion) can be realized:* For example, the payback period or time at which the project has paid back the owner's original investment falls in this category.
- *Variables that measure the difference between two or more compared schemes:* For example, this includes the savings–investment ratio used in life cycle cost analysis, which is a ratio comparing the savings achieved by one scheme over a second to the additional investment (cost) needed for the first. These measures also are called incremental measures
- *Criteria based on risk and uncertainty* (from decision theory and game theory): These are decision-making rules used with other performance measures, aimed at helping investors determine which of several schemes is the most preferable one when the schemes or solutions have different levels of performance under a number of different assumptions about what the future context will be.
- *"Feasibility" criteria:* These are not really measures of performance but a form of analysis that aims at determining whether the assumptions that must be made in order for the project to be competitive or to achieve a given expected performance are realistic assumptions, given the available market information.

Some of the main measures in each of these types are discussed below, except for straightforward costs and benefits, which were dealt with in previous chapters.

NET BENEFITS (BENEFITS MINUS COSTS)

Buildings are built for some purpose. Having that purpose met or satisfied is the benefit an owner derives from the building. In order to evaluate and make meaningful selections among proposed project alternatives, one must either work on the assumption that the benefits of all alternatives are equal, and then make the evaluation based on their respective costs, or be able to measure the benefit derived from the building in some other way.

For all those cases where the benefits of a project are flows of money, one obvious approach to the question of measurement is to use those amounts of money as the desired measure of performance, and to subtract from them the costs needed to achieve the benefits. The bottom line is what is left of money proceeds from a project after all the costs have been paid. Many widely used measures of performance fall into this category of net benefits or benefits minus costs. Distinctions among these measures depend on what kinds of costs are considered; so some conventions must be established concerning rules according to which the amounts are to be calculated, and then those rules must be applied consistently. For example, in rental properties, a starting point would be the potential gross income (PGI) or scheduled gross income—the hypothetical income a building could produce if there were no vacancies or bad debt losses. A series of net income measures results from the application of successive subtraction of costs from this:

- The Effective Gross Income (EGI) is the potential gross income minus the vacancy and bad debt losses (VAC):

$$EGI = PGI - VAC \qquad (7\text{-}1)$$

- The Net Operating Income (NOI) is calculated by

subtracting the annual operating expenses (EXP) from the effective gross revenues:

$$NOI = EGI - EXP \qquad (7\text{-}2)$$

- Cash Throw-Off (CTO) is the net operating income minus debt service (mortgage payments) (DS):

$$CTO = NOI - DS \qquad (7\text{-}3)$$

The CTO can be calculated for a given single year i (CTO_i) or averaged, or it can be cumulative (from the first year up to a specified given year).

- Taxable Income (TAXINC) is the remaining income on which income taxes are calculated, after subtraction of applicable deductions such as replacement reserves (REPRES):

$$TAXINC = CTO - REPRES \qquad (7\text{-}4)$$

- Spendable Cash After Taxes (SCAT) is the amount of money an investor has left over after having paid all expenses, debt service, and income taxes for a given year:

$$SCAT = TAXINC - INCTAX \qquad (7\text{-}5)$$

Again, SCAT can be estimated for a single year or averaged over a number of years. Figure 7-1 shows the succession of income-minus-cost categories discussed so far.

- The Sale Value (SALVAL) of the project at a specified time is the (estimated) amount of money the investor would obtain by selling it. This is also known as reversion.
- Net Sale Value (NETSALVAL) refers to the sale value minus the costs of selling it, that is, the realtor's fees and other costs associated with the transaction (SALCST):

$$NETSALV = SALVAL - SALCST \qquad (7\text{-}6)$$

- Project Net Worth (PNW) measures the value of the property at some specific point in time as established by subtracting the remaining outstanding mortgage balance (MORTBAL) from the Net Sale Value. It may or may not include the value of earlier profits from the project up to that time, that are reinvested and earning interest.

$$PNW_i = NETSALVAL_i - MORTBAL_i \qquad (7\text{-}7)$$

Both Net Sale Value and Project Net Worth can be looked at before or after tax, which is, in this case, the capital gains tax, if any.

- The total value of all net benefits (monetary or other) derived from the project over a specified period, usually measured in present worth equivalent, is the NETPWBEN:

$$NETPWBEN_n = \sum_{i=1,2}^{n} ((BEN_i - CST_i)/(1 + DISC)^i \qquad (7\text{-}8)$$

where BEN_i and CST_i refer to benefits and costs in year i, respectively.

- For the task of comparing the worthwhileness of sev-

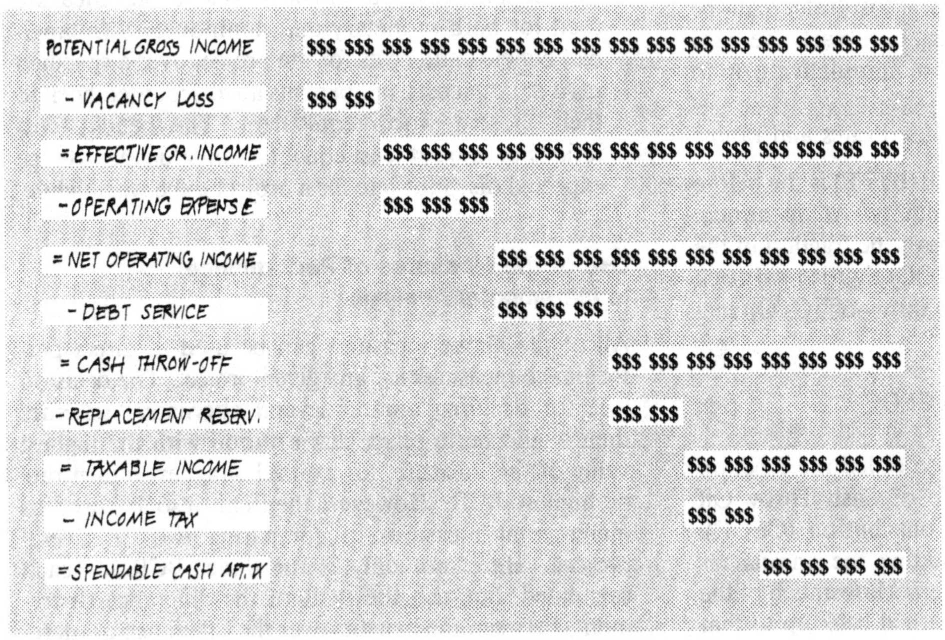

FIGURE 7-1.   Performance measures based on rental income minus costs.

eral alternatives, the difference in benefits or revenues (or any of the above measures) between one alternative and another can be a helpful measure of performance.

RATES AND RATIOS

The absolute dollar value of either costs or benefits, and even the difference between them, is of little use in comparing alternative projects, as the following example shows: Project A costs $1,000,000 and has a gross benefit of $1,001,000, that is, a net benefit of $1,000. Project B costs $1,000 and has a gross benefit of $2,000, that is, a net benefit of $1,000. Comparing their net benefit, one would say that they are equally preferable. However, it is intuitively obvious to most people that project B is a "better deal." A ratio of benefits over costs (BCR) would more clearly express this difference:

Project A: BCR(A) = 1,001,000/1,000,000 = 1.001
Project B: BCR(B) = 2,000/1,000 = 2.0

Thus ratios formed by dividing benefits by costs are more useful measures of performance in those situations where one of the variables—cost or benefit—is not fixed.

Rates are special ratio measures that express the benefit–cost ratio per time period—for example, the annual return per dollar invested, or the rate at which an investment grows each year. One often hears reference to the rate of return as a measure of performance. This concept answers the simple question: what share of the dollar invested "returns" to the investor each year? There are several methods of calculating it that must be carefully distinguished. Most rates can be computed for a single period (e.g., one year), averaged over a number of periods, or taken cumulatively up to the end of a specified period.

The following are commonly used rates and ratios:

- The Before Tax Equity Payback (BTEP) is the before tax return over the invested equity. As an annual measure, this return rate expresses the fraction of the initial investment (equity) (CEQC) that is paid back by the project through the cash throw-off (CTO$_i$) in a given single year i:

$$BTEP_i = CTO_i/CEQC \qquad (7-9)$$

This version of the BTEP is also called the Cash on Cash Rate (COCR)—the ratio of Cash Throw-Off (CTO) to the Cash Equity Contribution (CEQC). Computed cumulatively by adding the annual returns up to a given year, it expresses how much of the original investment has been earned back up to that point:

$$CUMBTEP_n = \left(\sum_i^n CTO_i \right) /CEQC \qquad (7-10)$$

- Cash return on total capital investment is the annual return CTO over the total initial project cost (TIPRJCST). This is sometimes called the Rate of Return (ROR) on total capital investment, as opposed to Return on Equity (ROE), which could refer to several of the return rates discussed here using equity in the denominator.

$$ROR = CTO_i/TIPRJCST \qquad (7-11)$$

- Yield (YLD) or After Tax Cash on Cash Rate is the Spendable Cash After Taxes (SCAT, including tax benefits, if any) over the equity:

$$YLD = SCAT/CEQC \qquad (7-12)$$

- The Internal Rate of Return (IRR) is that discount rate which makes the present worth of all annual net returns (NETBEN) and the net sale proceeds (reversion) equal to the initial equity contribution. It cannot be calculated directly but usually is found by iteration in the following equation:

$$CEQC = \sum_i^n (NETBEN_i/(1 + DISC)^i) \\ + (NETSALVAL_n/(1 + DISC)^n) \qquad (7-13)$$

- The Adjusted Internal Rate of Return (AIRR) or Modified Internal Rate of Return (MIRR) refers to an IRR in which the spendable cash after taxes has been reinvested year by year at some specified reinvestment rate, and therefore earns interest that constitutes part of the overall benefit of the project.
- The Benefit–Cost Ratio (BCR) is a general measure used for both monetary and nonmonetary (intangible) benefits and costs for the project over a specified period; both costs and benefits are discounted back to their present worth equivalent. The benefit–cost ratio should be at least equal to 1.00 plus the average expected discount rate for a project to be worthwhile.

Use of the Measures of Performance in Incremental Analysis

All of the above measures can be used as direct performance measures for any given project. The decision rule for deciding among several alternative schemes then would be to select that solution with the highest value of the measure, for example, highest return rate or highest BCR. However, they can also be used as incremental measures in comparing two or several projects. Here, not only is the direct BCR for each calculated, but also the BCR of the difference (increment) between the benefits over the difference or increment in cost.

- *Capitalization rate or cap rate:* This is the rate of return *on* an investment (in a loan, the interest rate) plus the rate of return *of* the investment (in a loan, the recapture rate or rate at which the principal returns to the lender). This is also called the Overall Rate (OAR). The mortgage cap rate is the rate at which the mortgage earns money—the annual mortgage payment divided by the loan amount. This is equivalent to the capital recovery factor or mortgage constant (see Appendix 8 and the conversion tables).
- *Net Income Multiplier:* The property cap rate is the rate at which a property earns money; its reciprocal is the Net Income Multiplier (NIM).
- *Default ratio:* Strictly speaking this is not one of the income/investment ratios as discussed above; it is that vacancy rate at and above which the revenues from a project no longer cover expenses and debt service. The project would then operate at a loss.

PERIODS

Determining the amount of time needed for a project to pay back the investment (or to break even) is an example of the use of a time period as a measure of performance. Other things being equal, investors prefer projects with shorter payback periods. Breakeven periods serve as a useful basis for comparison between alternative projects: how long does it take for the lower annual costs for a proposed project A to offset its higher initial costs as compared to alternative B?

- The Payback period (or simply payback) is a common measure referring to the amount of time required for the initial investment—usually the cash equity contribution—to be recovered. Distinctions are made between simple payback (SPB) and discounted payback. The former is found by dividing the CEQC by the first year's CTO:

$$SPB = CEQC/CTO \qquad (7\text{-}14)$$

It is a rather crude measure and does not account for the possibility that the CTO's may be unequal year by year-falling, rising, or otherwise fluctuating. The discounted payback does begin to account for this possibility; it is found by discounting each year's CTO_i to present value before adding up the CTO's and by doing this for as many years as necessary to reach the CEQC.
- The amortization period (term of a loan or mortgage) is the time in which the bank (lender) recovers its investment, the loan.
- Breakeven period sometimes refers to the time needed to recover the initial investment, which then would be the same as the payback period above; more commonly it means the time at which two project alternatives show the same performance,

using any of the other performance measures discussed.

PERFORMANCE MEASURES ACCOUNTING FOR RISK AND UNCERTAINTY: DECISION THEORY CRITERIA

The straightforward use of all the measures of performance discussed above is based on a tacit but unrealistic assumption that we can predict with certainty how a given solution will perform in the future. Our predictions are based on so many other assumptions that we make with only slight assurances that they will come true, that the result must be seen as having no more than a degree of probability.

There are many situations where one can foresee a number of different possible states of the world or distinct, mutually exclusive, context conditions. Because context co-determines performance, for each state of the world a given solution alternative must be expected to show a different level of performance. The states of the world have different probabilities of occurrence. Solution A may have the best performance if state-of-the-world X occurs, but B will be better if state-of-the-world Y happens. Then, the simple decision rule for certainty situations, of selecting the solution with the highest measure of performance, cannot be applied.

Decision theory and game theory are branches of mathematics that explore decision-making rules that could be used in such situations. They first distinguish three different kinds of decision-making situations, according to whether the states of the world can be predicted with certainty, or degrees of probability only (i.e., some degree of risk), or uncertainty (i.e., no reasonable forecast can be made about the probabilities of the different context situations). For each type of situation, criteria and decision-making rules have been developed that can be used in conjunction with some other measure of performance. Strictly speaking, the recommendations of decision theory involve several layers. At the base is some performance measure, which could be any one of the measures discussed above. This measure then is modified in some way, for example, by looking at the values of the measure given a number of context conditions or by applying the probability of context events to the performance measures, and so on. Finally, a decision rule is provided for making the selection. The recommended measures and rules are, briefly:

- *Expectation Value:* The expectation value E_A for a given solution alternative A, and a set of possible states of the world or context conditions, is the sum of outcomes O_{Ai} (the value of the performance measure achieved under each of the states of the world i), where each outcome O_{Ai} is multiplied by its probability P_i of occurrence:

$$E_A = \sum_i (O_{Ai} * P_i) \qquad (7\text{-}15)$$

P_i is a number between zero and 1; and the P_i's are adding up to one. The recommended decision rule is to choose that alternative from a set of competing solutions that has the highest (or lowest, if the performance measure is cost or some other measure to be minimized) expectation value.

- *The Min Max or Max Min rules*, depending on whether we are looking at measures of costs or benefits, respectively: "Minimize your maximum costs" —that is, choose that alternative for which the highest cost that can occur under any state of the world is the lowest among the alternatives; and "Maximize your minimum benefits"—choose that alternative for which the worst benefit given any context situation is the best among the alternatives.
- *Bayes' rule of equal likelihood for situations of uncertainty:* When probabilities for the states of the world cannot be established, use an expectation value computed with assumed equal probabilities for all alternatives.
- *Hurwicz alpha:* The economist Hurwicz proposed a "partial optimist/pessimist criterion" that allows the decision-maker to specify the degree of optimism or pessimism with which he or she views the situation at hand, and then to choose the alternative with the best expected performance for that optimism/pessimism degree.
- *The rule of minimizing one's maximum regret or opportunity cost:* This looks at the difference between the performance of an alternative and the best one could have done given the state of the world assumed as opposed to the one that actually occurred. The implied recommendation is to choose that alternative that involves the smallest regret.

The decision theory criteria and their use are discussed in more detail in Appendix 11.

CHOOSING THE MEASURE OF ECONOMIC PERFORMANCE

The preceding discussion presents a bewildering array of possible performance measures. Which measures should we use in a given situation? How do we go about making a reasonable selection? The following considerations may assist in this decision.

Available Information; Delivery Process Phase

One consideration that must be taken into account in choosing an appropriate measure of economic performance is the kind and reliability of available information (data), concerning both the project itself and its context, at the time when we need the analysis. This is a difficult problem, in spite of the huge amounts of data being collected and distributed. For one, the information available about the building project itself varies with the phase of the delivery process. During the feasibility analysis and programming stage, we do not know enough about the building itself because most decisions about the building have not yet been made. This lack of knowledge limits the kinds of measures that can be used to those that do not depend on such information. As the plans progress, both the detail and the nature of these decisions change, and then other measures become more relevant. So we must choose the measure of performance in view of the stage of the planning process at which we happen to be.

Quality of Data

The quality of available data about construction costs and continuing costs, let alone expected returns and so forth, varies considerably with the building type. Some building types have been studied very extensively, so that enough is known about costs, rental rates, operating costs and so on, whereas other buildings are unique one-of-a-kind projects for which it may be quite difficult to find adequate data. This means that we should select the measure of performance in view of the kind and detail of information available about the building type.

Cost of Data and Making the Analysis

Information is not a free commodity. It must be paid for in one way or another—either by paying some other company or agent for it, by assigning one's own employees to compile it, or by devoting time and energy on a continuing basis to the task of building up and maintaining an in-house collection of data over the years. For some analysis tasks, needed information may be readily available in standard format at low cost. For other projects, no pertinent data may be available at all, and then it could be quite expensive to compile what is needed on an ad hoc basis.

The cost of carrying out the required analysis must be added to the cost of data. With modern calculating equipment and the increasing connection between drafting and estimating software (CAD programs that immediately provide quantity takeoffs and cost estimates), the time and effort required for analysis will be considerably less than they used to be. (The cost of the computers and software still must be prorated in the cost for each project though, as must the cost of training personnel in their use.) Nevertheless, it is easy to see that there may be situations in which the added assurance a more detailed estimate would bring may not be worth the cost of obtaining and using the data needed to get a specific measure of performance. Thus, the choice of measure may be influenced by the cost of data and the analysis effort.

An interesting form of cost of information that usually is not considered is the cost of trial-and-error: the cost of mistakes we may have made because of not knowing better. They provide the information at the cost not of preventing but of repairing the consequences and damages. Although the information comes too late to benefit the project where the mistakes are made, it can be most valuable for future projects. Unfortunately, in our society where the reaction to mistakes is to place blame, there is little incentive (for those who know best, the people who made the mistake) to make the corresponding information available to others. I suspect that we collectively could do much better in using the information provided by the trial-and-error processes.

Calculating Equipment

A factor that in recent times has done much to reduce the cost of performing the needed analysis is the available calculating equipment, which was briefly mentioned in connection with the cost of data. Only a few decades ago, the choice of performance measure was not a big problem, mainly because it was too time-consuming and therefore expensive to perform the required calculations for more than the most basic measures. The development and the availability of programmable calculators and affordable computers have promoted the development and use of increasingly sophisticated performance measures. This capability also has made it possible to run performance analyses repeatedly for many alternatives, so as to systematically improve design solutions toward optimal performance,

PERFORMANCE MEASURE	INVESTOR / OWNER OBJECTIVE OR CONCERN					
	Profitable Sale "Now"	later	Occupy Initial cost	Long-term cost	Utility/ comfort	Rental income
COST (only) --			("Budget")	(Cost-in-use)		
INITIAL	no	no	Yes	no	n.a.	n.a.
LIFE CYCLE	no	yes until sale	?	Yes	n.a	n.a
BENEFITS (only)						
SALE PRICE	Yes	yes	n.a.	no	no	no
EFFECTIVE INC.	n.a.	?	n.a.	n.a.	n.a.	limited
EVALUATION SCORE	no	?	?	?	Yes	?
NET BENEFITS (B-C)						
NOI	no	until sale	n.a.	n.a.	n.a.	yes
CTO	no	until sale	n.a.	n.a.	n.a.	yes
TXINC	no	until sale	n.a.	n.a.	n.a.	yes
SCAT	no	until sale	n.a.	n.a.	n.a.	yes
NETSALVAL	Yes	YES	?	?	n.a.	yes
PNW	Yes	YES	?	?	n.a.	yes
DIFFERENCE	single pr.	several alt.				
INCREM. COST	n.a.	no	yes	yes	?	?
INCREM. BENEFIT	n.a.	yes	n.a.	n.a.	?	yes
INCREM.BCR	n.a.	yes	n.a	n.a	yes	yes
SIR	n.a.	yes	yes	yes	?	?
RATES & RATIOS						
BTEP	no	until sale	n.a.	n.a.	n.a	yes
COCR	no	until sale	n.a.	n.a.	n.a	yes
ROR	no	until sale	n.a.	n.a.	n.a	yes
YIELD	no	until sale	n.a.	n.a.	n.a	yes
IRR	no	until sale	n.a.	n.a.	n.a	yes
MIRR	no	until sale	n.a.	n.a.	n.a	yes
BCR	?	yes	?	yes	yes	yes
PERIODS						
BREAKEVEN	n.a.	yes	n.a.	yes	?	yes
BRKEVEN-COMPAR.	n.a.	yes	n.a.	yes	?	yes
DECISION THEORY						
EXPECTATION VALUE	n.a.	yes	yes	yes	yes	yes
MAXMIN / MINMAX	n.a.	yes	yes	yes	yes	yes
REGRET	n.a.	yes	yes	yes	yes	yes
BAYES	n.a.	yes	yes	yes	yes	yes
HURWICZ	n.a.	yes	yes	yes	yes	yes

FIGURE 7-2.   Applicability of economic performances measures to different conditions.

whereas in the "old days" one could afford perhaps one in-depth analysis for an otherwise already adopted solution to see if it would perform within an acceptable range.

Today, there is no such excuse for not routinely carrying out a number of analyses accompanying every phase of the programming, planning, design, detailing, and construction work. However, there are enough differences in the available hardware and software in a given situation that these differences will influence to some extent which measure of performance we use.

Investor Objectives

Finally, and most important, the choice of the measure of economic performance must be made in view of the owner's or the investor's concerns and economic objectives for the project. Some buildings are seen primarily as shelter for some activity or process (including residing in them), which is the real source of income or enjoyment. Receiving rental income from a building is the main purpose of the investment for some owners, whereas still other investors are looking primarily for the proceeds (profits) from resale. Whether these benefits are expected to be reaped over a long period or over a few years only, all these considerations call for different selections of measures of performance.

It makes a difference whether we wish to design a building to perform well in view of long-term performance or whether the short term is considered more important, and it is this relationship between design and economic performance that should be our main interest here. Unfortunately, it is this relationship about which our knowledge generally is most unreliable and often is based on little more than speculation.

Figure 7-2 is a table that lists the various measures and shows how they are likely to be applicable for various considerations discussed above.

STUDY QUESTIONS

1. Why is it generally not meaningful, when one is considering several design alternatives, to use either cost or benefit only as the deciding performance measure?
2. Why is it meaningful to use a benefit-minus-cost measure as the sole criterion for choosing among a set of design alternatives only when either costs or benefits are approximately the same for all those alternatives?
3. When is it advisable to use measures of performance that discount future benefits to present value before aggregating them into an overall measure (e.g., modified IRR or discounted payback)?
4. What measures of performance would be meaningful for the owner of a residence who will live in the home after his or her impending retirement, and who is concerned with the annual continuing utilities and upkeep costs?

8
Techniques of Economic Performance Analysis for Building Projects

OVERVIEW

In previous chapters we surveyed a number of economic measures of performance and the variables that go into each of them, along with some exploration of the needed data and their sources. It remains for us to assemble these measures into procedures for actually carrying out economic performance analysis. This chapter examines a selected number of such techniques and methods that have been described in the literature and used in practice, and provides step-by-step guidelines for their application.

The selection of approaches is not intended to be exhaustive. In some cases, the boundary between a genuine technique and mere measures is not entirely sharp, and the representation of the procedures also is genetic; that is, the circumstances of actual cases of application will call for considerable modification. The discussion will include Benefit–Cost Analysis, Incremental Analysis, Analysis of Payback, Life Cycle Cost Analysis, rate of return analysis, value engineering, Real Estate Financial Feasibility Analysis (Front Door and Back Door Approach), and the Real Estate Pro Forma Statement.

What Is a Technique?

It may be useful to review briefly what is meant by a technique in this context. The terms "technique," "process," "method," "procedure," and so on, often have been used interchangeably. A technique is simply a sequence of steps carried out for the purpose of solving some problem or achieving some task. The task can be described in terms of the variable (or set of variables) whose viability is to be examined. Embedded in a specific view of the problem situation, it proceeds from some starting assumptions or data—givens—toward an outcome that normally is expressed in terms of some measure of performance. It will be useful to keep these concepts in mind when we discuss individual techniques in the following presentation; some techniques have been named for the measure of performance used and others for the data needed, which could cause confusion. Figure 8-1 presents a table listing various common techniques in relation to their starting assumptions and outcome performance measure. The table might be modified somewhat in response to the nature of the task at hand in a given situation—whether it is primarily a question of making a decision (e.g., adopting or rejecting a proposed solution, plan, etc.), selecting the most appropriate one from a set of alternatives, or developing or designing the solution (program, design schematic, detailed solution, combination, etc.). The reader should try this modification as an exercise.

The table or matrix includes some procedures that either have been discussed in previous chapters, such as cost estimating techniques, or that should properly be labeled "design" or solution development.

The discussion of the techniques will follow the format of examining the nature of the task or problem to be solved, presenting the starting assumptions and needed data (givens), summarizing the sequence of steps, and discussing the outcome of the analysis and its use in making decisions.

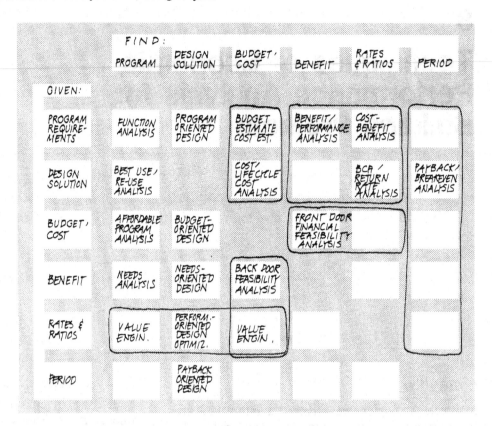

FIGURE 8-1. Techniques of economic analysis for building projects in view of givens and expected outcome of task.

BENEFIT–COST ANALYSIS

Concept and Aim of Benefit–Cost Analysis

Benefit–Cost Analysis is an example of a performance analysis technique. It can be described as a systematic, analytical approach to the evaluation of alternative courses of action or projects so as to identify the most advantageous one. Although the term refers to well-established procedures for evaluating the merits of public or government projects in particular, and much has been written about this approach, this explanation essentially represents what we have been trying to do all along in this book—to examine the economic benefits of building projects in some systematic fashion. In a more restricted sense, Benefit–Cost Analysis simply involves the use of a specific kind of measure, the Benefit–Cost Ratio (BCR), in the assessment of project alternatives. (See Chapter 7 for the BCR formula.)

The performance measure implies a recommended decision rule: adopt that alternative which shows the highest BCR; at the least, an acceptable alternative should have a BCR of 1.0 or better.

Benefit–Cost Analysis can be used as a straightforward economic analysis technique focusing on monetary variables only; but also—and this has been the main argument for its application to government projects—it can be used to include nonmonetary costs and benefits. It does this by assigning a money value to each intangible cost or benefit, and then including these values in the money amounts of the analysis.

Starting Assumptions/Givens

Benefit–Cost Analysis starts from a description of the proposed project (program or fully designed solution), the market conditions for cost estimates (construction prices, market value or purchase price of site, standard data regarding initial costs or services needed to complete the project, as well as cost-in-use assumptions), and the market conditions for benefits (rental income or sale prices). Assumptions must be made about the projected life span, timing of cyclical and noncyclical expenses and benefits, depreciation, inflation, and most of all, the discount rate used to convert all future costs to present worth. A controversial issue in the application of the technique to include intangible project costs and benefits invariably is the amount of money that should be assigned to such intangible aspects, especially if costs and benefits are not borne and enjoyed by the same individuals or groups in the same way.

Benefit–Cost Analysis: The Process

The technique of Benefit–Cost Analysis involves the following steps:

1. Clarify client objectives. Establish the discount rate to be used for converting future costs and benefits to a common time value. Identify all the alternatives to be compared, and ascertain their respective benefits and costs. This should include both monetary and intangible costs and benefits, to which an appropriate dollar value is assigned.
2. Examine the distribution of costs and benefits over time—draw a cash flow diagram for a better overview—and convert all cash flows to a common time basis. This could be either the present worth or the equivalent uniform annual payment.
3. Form the benefit–cost ratio (BCR or B/C) with the discounted cash flows.
4. Having repeated steps 1 through 3 for all alternatives, compare them and select the solution with the highest BCR, provided it is at least 1.0.

An additional, useful step before making the final selection would be to examine the specific cost and benefit elements of the alternatives to learn how they might be modified and improved to yield a better performance.

Problems Concerning the Benefit–Cost Ratio (BCR) and Its Use

In following this procedure, a few questions arise. Some of these are technical in nature, others more fundamental.

One such question has to do with intangibles. Accepting the idea that the value of intangible costs and benefits can be expressed in money terms, what dollar amounts should be assigned to the intangibles? How are they to be determined? Answers to this question range from the confident assertion that one can arrive at meaningful money amounts by expressing values, for example, through the process of negotiation or bargaining, to the recommendation of simply excluding the "irreducible" aspects from the analysis.

In the bargaining approach, one would, for example, establish the value placed by a person or group on some expected benefit by confronting the decision-maker with a very low cost he or she would have to incur to obtain the benefit, and then gradually increasing this cost until it is thought that the benefit is no longer worth the cost. The last acceptable cost is the monetary value that should be assigned to the benefit in question.

With respect to the answer that intangibles simply should not be included in the analysis, the very least that the architect should do is to point out what these intangible costs and benefits are and that they have been excluded. Otherwise, the analysis easily could be misinterpreted as having considered all relevant factors, and the resulting recommendation as needing no further thought or qualification.

A middle-ground approach involves complementing the actual Benefit–Cost analysis with a judgment-based evaluation closer to the evaluation process advocated by Musso and Rittel (in "Über das Messen der Guete von Gebaeuden" a chapter in a book *Arbeitsberichte zur Plannngsmethodik* 1 issued by the Institut fuer Grundlagen der Modernen Architektur, Stuttgart University; Karl Kraemer, Stuttgart 1968) and described in Chapter 4. For example, Riggs and West (in their book *Engineering Economics*) describe a set of criteria used in formal environmental quality evaluation procedures to evaluate intangible environmental resources such as open spaces, wilderness areas, and the like. These resources are rated on a judgment scale (1–10) according to quantity (e.g., size of areas), quality (e.g., desirability as compared with similar resources), human influence (e.g., the degree to which people would use a resource), uniqueness (e.g., relative frequency of occurrence of resources of the same kind), and significance (e.g., reversibility or irreversibility of damage to the resource, magnitude of adverse effects). It may be better in such circumstances to adopt this kind of procedure entirely for the evaluation of the alternatives, including their economic performance—translating everything including economic performance measures into judgment scores.

The next question concerns the rate at which future costs will be discounted to a common time basis, usually present worth. This question is of course encountered any time we have to make economic decisions involving comparisons of future and present cash flows; but it has been discussed with the greatest intensity and controversy in connection with Benefit–Cost Analysis, particularly with regard to public and government projects.

Which discount rate should be used? The recommendations for the choice of discount rate range from personal considerations of the rate of return for the best available alternative investment opportunity, which would be determined by each individual or firm (own personal discount rate), appearance of market interest rates, and the rate at which government borrows money from the private sector, to recommendations for a "social" discount rate that is lower than our individual discount rates. The choice of discount rate reflects our philosophy with regard to how much we as individuals, or we as a society (represented by government), should care for the future. To some advocates, the attitude of "consumption now" is expressed by high discount rates, as opposed to concern for the welfare of future generations, which would be best served if we made today's decisions using a low discount rate. For example, planting oak trees will not be seen as making much sense economically if the benefits of having a large oak tree, expected to mature in a few hundred years, are discounted with a high discount rate; its present worth will be too small to justify the expense and effort. Thus, no general recommendations can be given here to alleviate the problem of the choice of discount rate for each new project.

Another question is referred to as the issue of displacement of costs and benefits: Are "disvalues" costs or reduced benefits? This question concerns certain items that can be seen either as reduction of benefits or as costs. It is therefore not clear whether they should be accounted for in the numerator or in the denominator of the B/C formula.

For example, consider a planned commercial office building. One alternative involves the design of the building for the specific requirements of a single corporate tenant (which would guarantee the lease for a specific number of years in return for certain design modifications); another would be aimed at many smaller tenants. The second alternative would entail vacancy losses that would not occur in the first scenario. In comparing the two schemes, should the vacancy losses be counted as reduction of income, or as costs? This decision could make a significant difference in the BCR: Assume annual gross revenues to be $200,000 and annual costs to be $150,000 for the first alternative, resulting in a BCR of $2:1.5 = 1.33$. Everything remaining the same, if a vacancy loss of 10% is introduced, this would lead to a BCR of $(200,000 - 20,000)/150,000 = 1.2$ if the vacancy loss is counted as reduced benefit, but to a BCR of $200,000/(150,000 + 20,000) = 1.18$ if it is considered an added cost.

In the comparison of alternative solutions, the order of preference does not change as long as the BCR is calculated the same way for all solutions. However, there could be difficulties in interpretation if several solutions featured different kinds of such adjustments of costs and benefits, or if some solutions had adjustments and others did not have them. There is no general rule for avoiding or remedying such problems. The only recommendations that can be made are that, first, issues of this kind should be pointed out to the client for whom an analysis is being done, so that differences in how the problem is viewed at least can be discussed; and, second, in comparing a number of alternatives, the assumptions regarding such displacement of costs and benefits must be treated in the same way across the board for all solutions. Benefit–cost analysis and similar techniques have been criticized precisely because the kind of manipulation is possible that can be used to make any preconceived alternative look good and others bad, because the manipulation can be hidden merely by not drawing attention to it (after all, the figures used are all correct), and because it can be explained away as an unintentional "honest" mistake if it is discovered.

Figure 8-2 gives a spreadsheet example of a benefit–cost analysis applied to the familiar office building schematic whose initial costs were examined in Chapters 3 and 4, and whose cost-in-use was explored in Chapter 5. In this version, the net leasable area is 60,000 sf. The Benefit–Cost Ratio is formed with the present worth equivalents of cumulative gross present worth benefits including sale value, over cumulative present worth costs plus equity investment; only the monetary costs and benefits before taxes are considered. It can be seen that in this example it takes a few years before the BCR reaches 1.0, and in 20 years it never becomes more than 1.39 — not a very impressive performance.

INCREMENTAL ANALYSIS

When several alternative solutions that must be evaluated are reasonably close together in their expected cost, the evaluation can use any of the measures of performance surveyed in Chapter 7 to arrive at the "best" choice. But alternative schemes often involve very different levels of initial investment, sometimes close to a level where the availability of investment money becomes the constraint determining the outcome, rather than the performance of the solutions. However, this situation raises another interesting question: Is the difference in performance (say, return on investment) between solutions A and B worth the difference in cost between them? Could the cost difference have been allocated to some other investment venture with even better results?

All of the measures discussed in Chapter 7 can be used as direct performance measures for a given project. The decision rule used to decide among several alternative schemes is to select that solution with the highest value of the measure, for example, the highest return rate or the highest BCR. But they can also be used as "incremental" measures in comparing two or several projects. Here, not only the direct BCR for each is calculated (using BCR as the example), but also the BCR of the difference (increment) between the benefits over the difference or increment in cost.

For example, consider the four project alternatives shown in Figure 8-3. Here we see that although all the projects have a worthwhile BCR of at least 1.11 — that of solution D, which also has the highest absolute benefit — solution A shows the best BCR, 2.0. However, in comparing the incremental benefit from any of the other projects with the needed additional cost, this added investment only pays off for project C, with an incremental BCR of 1.6, whereas both B and D remain at an incremental BCR of 1.0. In comparing each project with its next lower alternative, C shows a good BCR on the added investment of 3.0, whereas the added cost for D results in a not worthwhile BCR of 0.5 (each additional dollar invested only produces $0.50 in return).

EXAMPLE: SPREADSHEET FOR OFFICE BUILDING SCHEME A   (DOUBLE-LOADED CORRIDOR)

Input variables: Design variable values show   **bold**

Name	Symbol	Value	Unit of measurement	Source; Constraint	
Net leasable area	NLA		60000 sf	Program	
Module area	MAR		600 sf	Program	
Module length	ML	**20** lf		Design	ML ≥ 10'
Corridor width	CW	**6** lf		Design	CW ≥ 6'
Wall thickness	WT		0.8 lf	Design	
Building construction price	CPRC		52 $/sf of TFA	Context	
No. modules per floor	NMPF	**10** #		Design	
Floor-to-floor height	FHT	**10** lf		Design	

Equations:

	Variables	Symbol		Value Unit	Formula
8)	Module width	MW	=	30.00 lf	MAR / ML
7)	Number of modules	NOM	=	100 sf	NLA / MAR
6)	Number of floors	NOF	=	10	NOM / NMPF
5)	Building length	BL	=	211.60 lf	2*WT + MW*(NMPF + 4)/2
4)	Building width	BW	=	49.2 lf	2 * (ML + 2 * WT) + CW
3)	Floor Area (one floor)	FA	=	10410.72 sf	BW * BL
2)	Total Floor area	TFA	=	104107.20 sf	FA * NOF
1)	Building Construction cost	BLDCST	=	$5,413,574	TFA * CPRC

(Building construction cost is assumed to include tenant improvement costs)

9)	Net to Gross Ratio	NGR	=	0.58	NLA/TFA

Total Initial Project Cost	TIPRJCST	PW$	$8,064,963	
Equity investment	CEQC	PW$	$1,574,338	
Debt service (mortgage paymen	DS	A$	$802,912	
Potential gross income year 1	POTREV	PW$	$1,560,000	
Vacancy rate	VR	%	8.00%	
Vacancy loss	VLOSS		$124,800	VR*POTREV
Effective annual revenue (EGI)	ANREV		$1,435,200	
Annual expenses	ANEXP		$507,187	
Net Operating Income	NOI		$928,013	
Annual inflation rate	INFL	%	6.00%	
Annual site appreciation rate	SAPPR	%	5.00%	
Discount rate	DISC	%	8.00%	
Building life span	S	years	45	

FIGURE 8-2A.   Benefit–cost analysis—example.

	Year i	1	2	3	4	5	6	7	8	9	10	11	12	13	14	15	16	17	18	19	20
POTREVi	Gross revenues (in 1000's) = POTREV(1)*(1+INFL)^i	$1,560	$1,753	$1,858	$1,969	$2,088	$2,213	$2,346	$2,486	$2,636	$2,794	$2,961	$3,139	$3,327	$3,527	$3,739	$3,963	$4,201	$4,453	$4,720	$5,003
VLOSS(i)	Vacancy loss (in 1000's) = POTREV(i)*VR	$125	$140	$140	$140	$140	$140	$140	$140	$140	$140	$140	$140	$140	$140	$140	$140	$140	$140	$140	$140
ANREV(i)	Annual effective revenues = POTREV(i) - VLOSS(i)	$1,435	$1,613	$1,718	$1,829	$1,947	$2,073	$2,205	$2,346	$2,495	$2,653	$2,821	$2,999	$3,187	$3,387	$3,598	$3,823	$4,061	$4,313	$4,580	$4,863
ANEXP(i)	Annual expenses (in 1000's) = ANEXP(1)*(1+INFL)^i	$507	$570	$604	$640	$679	$719	$763	$808	$857	$908	$963	$1,021	$1,082	$1,147	$1,216	$1,288	$1,366	$1,448	$1,535	$1,627
NOI(i)	Net Operating Income (1000's) = ANREV(i) - ANEXP(i)	$928	$1,043	$1,114	$1,189	$1,269	$1,353	$1,443	$1,538	$1,638	$1,745	$1,858	$1,978	$2,105	$2,240	$2,383	$2,534	$2,695	$2,865	$3,045	$3,236
DS	Debt service (i) (in 1000's)	$803	$803	$803	$803	$803	$803	$803	$803	$803	$803	$803	$803	$803	$803	$803	$803	$803	$803	$803	$803
CTO(i)	Cash Throw-off(i) (1000's) = NOI(i) - DS	$125	$240	$311	$386	$466	$550	$640	$735	$836	$942	$1,055	$1,175	$1,302	$1,437	$1,580	$1,731	$1,892	$2,062	$2,242	$2,433
SVAL(i)	Site value (i) (1000's) = SCST *(1+SAPPR)^i	$500	$551	$579	$608	$638	$670	$704	$739	$776	$814	$855	$898	$943	$990	$1,039	$1,091	$1,146	$1,203	$1,263	$1,327
BVAL(i)	Building value (i) (1000's) = (TIPRJCST-SCST)*(S-i)/S	$7,565	$7,229	$7,061	$6,893	$6,724	$6,556	$6,388	$6,220	$6,052	$5,884	$5,716	$5,548	$5,380	$5,211	$5,043	$4,875	$4,707	$4,539	$4,371	$4,203
SALVAL(i)	Project sale value(i) (1000's) = BVAL(i)+SVAL(i)	$8,065	$7,780	$7,639	$7,500	$7,363	$7,226	$7,092	$6,959	$6,828	$6,698	$6,571	$6,446	$6,322	$6,201	$6,083	$5,967	$5,853	$5,742	$5,634	$5,529
SALCST(i)	Sale cost(i) (1000's) = SALVAL(i) * 0.06	$484	$467	$458	$450	$442	$434	$426	$418	$410	$402	$394	$387	$379	$372	$365	$358	$351	$345	$338	$332
MORTBAL(i)	Mortgage balance (in 1000's) DS*((1+MLINT)^(MP-i)-1)/(MLINT*(1+MLINT)^(MP-i))	$6,250	$6,197	$6,138	$6,072	$5,997	$5,914	$5,821	$5,716	$5,600	$5,469	$5,322	$5,158	$4,974	$4,767	$4,537	$4,278	$3,989	$3,664	$3,301	$2,894
NETSALVAL(i)	Net Sale Value (1000's = SALVAL(i) - MORTBAL(i)-SALCST(i)	1,331	1,116	1,043	979	923	879	845	825	818	828	855	901	969	1,062	1,181	1,331	1,513	1,733	1,995	2,303
PWANREV(i)	PW Effective income = ANREV(i) / (1+DISC)^i	$1,329	$1,383	$1,364	$1,345	$1,325	$1,306	$1,287	$1,268	$1,248	$1,229	$1,210	$1,191	$1,172	$1,153	$1,134	$1,116	$1,097	$1,079	$1,061	$1,043
PWANCST(i)	PW of Costs (i) (1000's) =(ANEXP(i)+DS)/(1+DISC)^i	$1,213	$1,177	$1,117	$1,061	$1,008	$959	$913	$871	$830	$793	$757	$724	$693	$664	$636	$610	$586	$563	$542	$521
CUMPWANREV(i)	Cum. PW benefits up to yr i =SUM(PWANREVi)	$1,329	$2,711	$4,075	$5,420	$6,745	$8,051	$9,338	$10,606	$11,854	$13,083	$14,293	$15,484	$16,656	$17,809	$18,943	$20,059	$21,156	$22,235	$23,297	$24,340
CUMPWBEN(i) if sold year i	Cum. PW benefits(i) = CUMPWANREV(i)+(NETSALVAL/(1+DISC)^i)	$2,544	$3,668	$4,903	$6,139	$7,373	$8,605	$9,831	$11,051	$12,263	$13,466	$14,659	$15,842	$17,012	$18,170	$19,315	$20,447	$21,565	$22,669	$23,759	$24,834
CUMPWANCST(i)	Cumulative PW Costs yr i = SUM(PWANCST(i))	$1,213	$2,390	$3,507	$4,568	$5,576	$6,535	$7,449	$8,319	$9,150	$9,942	$10,700	$11,424	$12,117	$12,781	$13,417	$14,027	$14,613	$15,177	$15,718	$16,239
CUMPWCST(i) if sold year i	Cumul. PW Costs(i) =CEQC+CUMPWANCST(i)	$2,787	$3,964	$5,081	$6,142	$7,150	$8,110	$9,023	$9,894	$10,724	$11,517	$12,274	$12,998	$13,691	$14,355	$14,991	$15,602	$16,188	$16,751	$17,293	$17,814
BCR(i)	Benefit-Cost Ratio year i	0.91	0.93	0.96	1.00	1.03	1.06	1.09	1.12	1.14	1.17	1.19	1.22	1.24	1.27	1.29	1.31	1.33	1.35	1.37	1.39

FIGURE 8-2B.   Benefit–cost analysis—example. Calculations for 20-year period.

Project: >>>	A	B	C	D
Benefits:	2,000	5,000	8,000	10,000
Costs:	1,000	4,000	5,000	9,000
B/C or BCR:	2	1.25	1.6	1.111
Difference $(B_i - B_A)/(C_i - C_A)$ (any project compared to the first)	-	$\frac{3,000}{3,000} = 1$	$\frac{6,000}{4,000} = 1.5$	$\frac{8,000}{8,000} = 1$
Difference $(B_{i+1} - B_i)/(C_{i+1} - C_i)$ (any project compared to the next higher one)		$\frac{3,000}{3,000} = 1$	$\frac{3,000}{1,000} = 3$	$\frac{2,000}{4,000} = 0.5$

FIGURE 8-3.   Incremental analysis—example.

This type of problem is the subject of incremental analysis of costs and benefits in alternative solutions. This technique often is discussed and used in connection with Benefit–Cost Analysis. Basically, incremental analysis applies a chosen performance measure (e.g., B–C or B/C) to the differences or increments between the costs and benefits of alternative solutions.

An example will explain this best. Consider a set of different possible solutions for a mixed-use project in an urban location in which the zoning regulations limit the FAR (Floor Area Ratio) and thus the amount of building that can be put onto the site. The FAR limits are: 1:2 for the first three stories, 1:1.5 for the fourth through tenth stories, and 1:1 above ten stories. Of course, the site is limited as well, and one of the crucial decisions is whether to try to accommodate parking on grade (in which case the parking will limit the number of square feet than can be served) or to build a parking garage in connection with the building. In the latter case, not only can the allowable FAR be fully exploited, but the zoning ordinance also gives an additional incentive, in that the parking structure only counts 50% toward the FAR. Furthermore, there is another incentive provided in the code, that for each square foot of retail area provided, an additional amount of total floor area will be permitted. A design strategy of trying to exploit the maximum FAR at each level of these rules will produce a total of four alternative solutions, each with a different cost and each with a different structure and level of returns (benefits).

Figure 8-4A presents the schematic design alternatives with their TFA, number of floor and FAR. Figure 8-4B shows a spreadsheet calculating the costs and benefits for each of these solutions. In addition, the differences between each solution and the next more expensive one are shown. Solution A is the cheapest one, a three-story building with all parking on grade, an annual cash flow of $14,119, and a BCR of 1.21. The other solutions all aim at exploiting a higher allowable FAR, which must be done at the expense of providing covered parking under the building. With covered parking, a higher rent can be charged; but with the assumptions as stated, none of the high-rise solutions approaches the cash flow or the BCR of the first solution. However, if a high-rise solution is to be used, the jump between solutions B and C, with a BCR of 1.45 for the added expense over solution B, suggests that trying for the incentive for retail space is worthwhile. From a pure cash flow point of view, none of the high-rise solutions in this example, given the stated assumptions, should be considered unless the rentals can be raised more substantially. In every case, the added investment over solution A will lose money, having a BCR under 1, even though seen in isolation all solutions have an acceptable (if unimpressive) BCR.

Note that in this case the incremental analysis has been done by using a Benefit–Cost Ratio based on annual costs and benefits plus initial costs distributed evenly over the 25 years of the mortgage term and ignoring inflation, which might have raised both rents and annual costs but not debt service. Thus inflation actually would make the solutions slightly more attractive.

The Incremental Analysis Process

1. Describe the alternatives to be compared.
2. Select a measure of performance on the basis of client objectives (e.g., Benefit–Cost Ratio). Establish the threshold level of performance that should be achieved, not only by each solution but also by the increment of cost between solutions. (For the Benefit–Cost Ratio, this would be 1 or 1 plus the investor's minimum attractive rate of return for alternative investments.)
3. Establish the performance of all alternatives with respect to the chosen measure.
4. Arrange the alternatives in order of the amount of investment or cost required.
5. Eliminate all solutions from consideration whose performance is lower than the next lower cost solution, respectively.

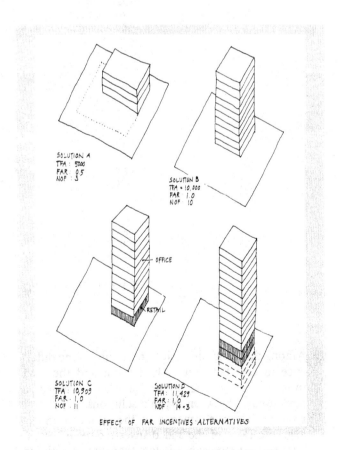

FIGURE 8-4A.   Incremental analysis—design alternatives.

INCREMENTAL ANALYSIS - EXAMPLE
MIXED USE PROJECT

Site area 100*100	10000	10000	10000	10000
Landscaping requireme	1000	1000	1000	1000
Parking requ. (sfNLA/s	200	200	200	200
Parking space requ(sf/s	280	280	280	280

	SOLUTION A	SOLUTION B	SOLUTION C	SOLUTION D
Office floors	3	10	10	12
Retail floors	0	0	1	2
Total floors	3	10	11	14
FAR	0.5	1	1	1
Incentive allowance	0	0	909	1429
TFA allowed	5000	10000	10909	11429
FA	1667	1000	992	816
Remain. site area	7333	8000	8008	8184
Net to Gross ratio NGR	0.8	0.8	0.8	0.8
Net leasable area NLA	4000	8000	8727	9143
Parking spaces required	20	40	44	46
Parking area required	5600	11200	12218	12800
Parking on grade	5600	0 8000	0 8008	0 8184
Undergr. parking	0	3200	4210	4616
Parking share of TFA	0	1600	2105	2308
Adjusted TFA for renta	5000	8400	8804	9120
Adjusted NLA	4000	6720	7043	7296
On grade parking - sf	5	5	5	3
Undergr. parking - sf	35	35	35	35
On-grade parking - cos	$28,000	$40,000	$40,041	$24,551
Undergr. Parking- cost	$0	$112,000	$147,347	$161,571
Building sf price	55	55	55	55
Building constr. cost	$275,000	$550,000	$600,000	$628,571
Total cost - bld. + park	$303,000	$702,000	$787,388	$814,694
Rental retail	25	25	25	25
Rental office	20	22	24	24
Revenue Retail	$0	$0	$24,793	$40,816
Revenue Office	$80,000	$147,840	$145,238	$135,928
Total Rev.	$80,000	$147,840	$170,031	$176,744
Mortgage term	25	25	25	25
Loan interest rate	0.1	0.1	0.1	0.1
Cap. recov. factor	0.1101681	0.1101681	0.1101681	0.1101681
Debt serv. (100% fin.)	$33,381	$77,338	$86,745	$89,753
Annual cost rate	$6.50	$6.50	$6.50	$6.50
Annual cost	$32,500	$65,000	$70,909	$74,286
Total annual cost	$65,881	$142,338	$157,654	$164,039

		DifferenceA-B		Difference B-C		DifferenceC-D	
B	$80,000	$67,840	$147,840	$22,191	$170,031	$6,713	$176,744
C	$65,881	$76,457	$142,338	$15,316	$157,654	$6,385	$164,039
B-C	$14,119	($8,617)	$5,502	$6,875	$12,377	$329	$12,706
B/C	1.21	0.89	1.04	1.45	1.08	1.05	1.08

	Difference A-C	Difference B-D	Difference A-D
B	$90,031	$28,904	$96,744
C	$91,773	$21,701	$98,158
B-C	($1,742)	$7,204	($1,414)
B/C	0.98	1.33	0.99

FIGURE 8-4B.   Incremental analysis—spreadsheet results.

6. Among the remaining solutions, establish the difference in cost between each solution and the next more expensive one, as well as the difference in performance between the two solutions.

7. Examine the performance of the cost increments (e.g., the BCR of the added investment). If the incremental performance is less than the established minimum performance threshold, the question should be raised of whether the cost difference should be invested in a different way that would ensure its achieving at least the minimum performance—even if the performance achieved by the more expensive solution itself still is above the threshold.

ANALYSIS OF PAYBACK; BREAKEVEN PERIOD

A common concern for the investor in a building project is how long it will take for the project to return or to pay back the initial investment. The time to achieve this, the payback period or simply payback, is considered a measure of performance: the shorter the time, the better the investment. Until that time, the investor is "in the red"; from then on, the project begins to earn money. This also is referred to as the point in time when the project breaks even or "has paid for itself." Of course, this is a reliable measure of performance only under certain simplifying assumptions—for example, if the cash flows are relatively stable and consistent over time. Otherwise, the distribution of wildly fluctuating payments and expenses (e.g., due to large repair and replacement bills at irregular intervals) may call for a closer analysis of how the cash flow pattern compares with alternative solutions, for example, in terms of present worth. There are several significant problems involved with the use of payback to make important building investment decisions. The technique is included here mainly because it still is quite frequently used, and to emphasize the problems.

The concept, use, and shortcomings of payback, the payback comparison method or payout method, the related notion of the breakeven period, and breakeven analysis and its various meanings are discussed in this section. The discussion assumes familiarity with some of the other common measures of economic performance such as cash throw-off or cash flow, and the concept of discounting future amounts to present value for comparison of different payment patterns (though the simple payback formula avoids consideration of the time value of money and therefore does not use discounting).

The Payback Concept

The concept of payback applies to situations in which an initial investment is needed to implement a project, that then will result in some form of annual income over time. The payback period is the amount of time needed for the annual net income to become equal to the initial investment—to "pay for itself." In its simplest version, the general formula for this is:

Payback period = Initial cost or
investment/Annual net income.   (8-1)

For example, if a project requires a $100,000 initial investment and produces a net income of $20,000 each year, its payback period or payback is five years. In this simple and most popular form, simple payback, there is no consideration of the time value of money; that is, the future value of the annual income flows are not discounted to their present worth equivalent. It is assumed that the life span of the project is longer than the payback period because the expectation is, of course, that it will begin to earn profit over and above the initial investment after the payback date.

The payback period then can be used to compare different project alternatives; the shorter the payback period, the better the project (i.e., the sooner it will begin to make a profit for the investor). This statement must be taken with a bit of caution; a fair comparison with payback can be obtained only if the payments are the same (or approximately the same) every year, and the life spans are similar for all compared proposals. Figure 8-5 shows three alternative cash flow pictures that have the same payback period. The method cannot distinguish between them although there obviously are some significant differences between them—especially if one considers how each alternative is likely to continue making money after the payback period. In fact, if the method uses the first year's income in the above formula, it will produce quite misleading results. The formula will yield the true payback period only for solution B, whereas solution A will appear to have a much shorter period, and C will seem to have an unacceptably long payback period. Actually, if the cash flows continued as shown, solution C would be the best alternative by far.

Figure 8-6 shows three alternatives with different initial costs and uniform incomes but the same payback period. It seems obvious that solution B, the most expensive one in terms of initial investment, is the one to be recommended; and solution C, which may look attractive to the unwary because of its low initial cost, is the least desirable. But, again, the method in its simple form does not distinguish between them.

Most measures of performance should not be used in isolation but, if possible, should be used along with other measures. This is especially true for payback; the above examples may be sufficient to emphasize this message. Some analysis recommended against using the payback criterion at all, whereas others recommend at least a year-by-year analysis to acknowledge differences in annual performance, and then to discount the annual cash flows to present value (discounted payback). The best solution then would be that which has the shortest payback period with discounted cash flows.

Breakeven Analysis

The concept of breakeven often is used in connection with payback or even as a synonym for it. Breaking even on the one hand means simply having earned the investment back. Its more important use is in the comparison of performance of several alternatives, especially when performance is distributed unevenly over the study period, and one alternative is initially less

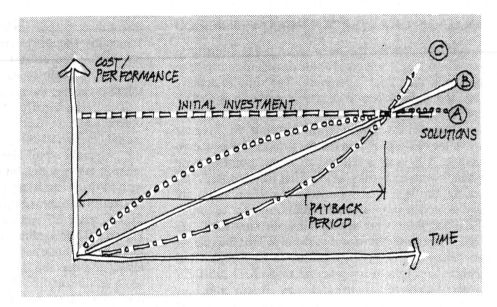

FIGURE 8-5. Payback analysis.

advantageous but improves later on, with another showing better performance in early years but declining later.

Breakeven analysis thus is not used as a performance measure but is used together with some other measure. The key question is: at what point in time is the performance of two alternative solution equivalent? This is significant for investors with a specific planning horizon (e.g., one's anticipated retirement date, or a time at which a new element of infrastructure, such as a new freeway or bridge, will be completed that then changes the parameters for the use of land in a given area). They would like to know which of several alternatives shows the best performance up to a specified point in time. This is also the typical form of analysis for common problem such as comparing solutions with different initial costs and different long-term costs: "If we choose the more expensive solution A, we will have lower

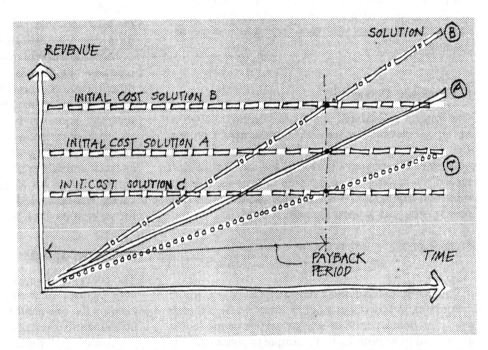

FIGURE 8-6. Payback analysis. Alternatives with different initial costs, uniform cash flows, and same payback period.

annual maintenance costs than for the initially cheaper solution B." The graphic representation of such problems, called a breakeven chart, is shown in Figure 8-7.

If cost only is the issue, a good measure of performance to use here is the cumulative total cost, preferably with annual costs discounted to account for the time value of money. There are two possible situations for such problems: (a) the two curves of cumulative discounted total cost intersect after some years; (b) they diverge or extend in parallel throughout the study period. If they intersect, as in Figure 8-7, the crossover point occurs when the alternatives break even. Before that point, if the performance measure is cost, solution A is preferable; afterward solution B has better performance. The choice of solution depends on the investor's planning horizon, and on whether the planning horizon is shorter or longer than the breakeven period; for investor X, solution A is preferable, whereas investor Y would prefer solution B.

If the performance curves do not intersect, one of the solutions "dominates" the other throughout the life span, and the decision task is trivial.

When several alternative solutions are compared, the breakeven charts will show a bundle of lines or curves potentially intersecting in many points. The interesting breakeven points always are found on the upper or lower boundary of such envelopes of curves: on the lower boundary for costs and similar criteria to be minimized; and on the upper boundary for returns, benefits, and other criteria to be maximized. Alternatives represented by lines or curves that do not touch the critical boundary are said to be dominated by the

other solutions and should not be considered. In Figure 8-8A, solutions A, B, and C all are candidates for the lowest cumulative cost solution, depending on the client's planning horizon. Solution D is dominated by the others and should be disregarded. Figure 8-8B shows the analogous situation for cumulative returns.

LIFE CYCLE COST ANALYSIS

The concept of life cycle cost as a performance measure was introduced in Chapter 7, and the basic principle of life cycle cost analysis explained. We must clarify the working assumptions and the specific steps involved in carrying out the analysis. Life cycle cost analysis can be carried out at very different levels of detail—in analogy to the area method, enclosure method, and unit price initial cost estimates. In each case, the cost estimates will have to be based on unit price assumptions appropriate to the level of information available at the respective stage of the process.

Assumptions for Life Cycle Cost Analysis

The purpose of life cycle cost analysis (LCCA) usually is to compare different competing solutions. The typical situation for the application of LCCA is that of deciding which of several subsystems involving energy should be installed, such as heating, cooling, and air-conditioning systems. For example, innovative energy-conserving technologies often carry a higher initial cost than conventional systems, and it must be determined whether the long-term savings are worth the higher

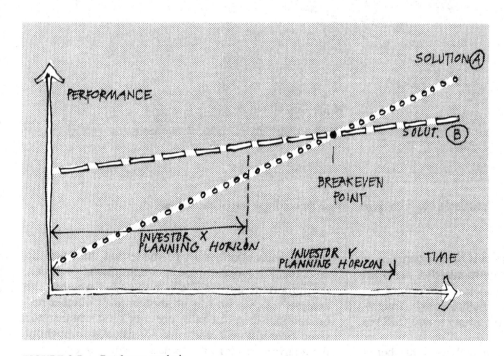

FIGURE 8-7.   Breakeven analysis.

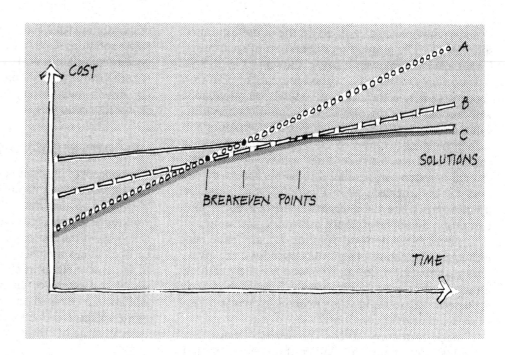

FIGURE 8-8A.   Breakeven analysis for several alternatives. The measure of performance is cost; the lower boundary is critical.

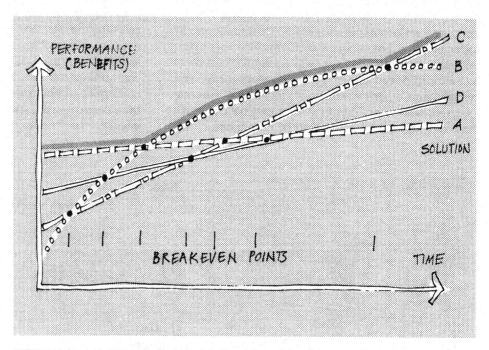

FIGURE 8-8B.   Breakeven analysis for benefits of several alternatives.

cost. Even when only one system is being examined, it must be compared against some standard of performance. The nature of the analysis task determines the measure of performance that should be used: straightforward LCC or the savings–investment ratio (SIR) as well.

In carrying out a life cycle cost analysis, a number of critical assumptions must be made that influence the outcome. For most of these assumptions and decisions, the important question is not so much whether the analysis is done one way or another (e.g., whether real or nominal discount rates are used) but whether the same assumptions are applied consistently throughout the analysis and for all alternatives compared.

Study Period. Once the alternatives to be compared have been identified and described, the first assumption is that of the study period. Most often, the study period will be the assumed life span of the building or of the subsystem in question. But other study periods can be considered; this planning horizon reflects the length of time that the investor expects to be involved with the project.

There are both statistics and standard assumptions about the life expectancy of ordinary buildings, which are reflected in tax guidelines for depreciation allowances. Of course, this information may start a self-perpetuating cycle, as the tax guidelines become the target life spans for which buildings are designed. This may not be a problem; but if one looks at the life span assumptions—for example, the AIA tables (in the AIA's *Life Cycle Cost Analysis*), based on Internal Revenue Service guidelines—one cannot help noticing that the periods in question are all well below the average person's life expectancy; for example, office buildings have a projected useful life of 45 years. However, some subsystems of the building must be designed with a considerable safety factor (e.g., the structural system), and these subsystems will have a much longer potential life span than 45 years; so it may seem strange to deliberately aim for such a short period. (One wonders what would have become of architecture if the builders of the world's great architectural monuments had taken a similar attitude.) These assumptions reflect the fact that upgrading facilities to keep their income-earning power intact becomes more expensive as time goes by, and it eventually will be cheaper to replace the building than to upgrade it.

In specifying the study period one must include the starting date or baseline date. This is the "present" to which all future costs are discounted. It is also the date to which sunk costs—those incurred prior to the baseline date, such as site costs and site improvements—must be brought forward, so that the analysis can proceed from an assumption of today's costs across the board.

Discount Rate. The second major factor that must be specified is the discount rate used to make conversions from future worth to present worth or annuities, as the case may be. There is no general rule about what discount rate should be used, except that it must reflect the investor's average expected rate of return on the best alternative investment opportunity and be stated in nominal terms; that is, the rate should include general inflation. For example, if the investor's expected rate of return were 10% assuming no inflation (the real rate) and inflation is in fact 6% annually, the nominal rate would be 16%.

Present Worth or EUAC Comparison. The next decision concerns whether the analysis will be done by using the present worth method or the equivalent uniform annual cost (EUAC) comparison. Either one will provide a valid basis for comparison. In some cases, it may be more important for an owner to see how the total costs relate to the ability of the project to generate the income needed to pay for them, on a year-by-year basis. This would be better shown by the EUAC than by the Present Worth method. On the other hand, if the purpose of the analysis is to study the relationship and trade-off possibilities between initial investment costs and long-term costs, the Present Worth approach might be preferred.

Current or Constant Dollars. A major issue in life cycle cost analysis concerns how inflation is dealt with. If future costs are quoted in their actual numerical value according to how much we predict those items will actually cost in a given future year, they are stated in "current dollars." If costs including future costs are quoted in terms of their cost at a specific point in time, say 1992, regardless of inflation, they are quoted in "constant dollars." Constant dollars always must be stated with the reference year. Quoting costs in constant dollars simplifies the analysis because it eliminates the steps of first applying inflation to find future values and then discounting these values back to present value for comparison. (If constant dollars are used, discounting must be done with the real instead of the nominal discount rate.)

Comprehensiveness. Finally, decisions must be made about the scope and comprehensiveness of the analysis. In principle, LCCA can be applied to a single element within a building as well as to the entire project. The level of detail will have implications for the time that will be needed and for the cost of the analysis; thus it is important to have a clear picture of the purpose and objectives of the analysis, so that the choice of system (subsystem) elements and therefore of significant cost elements will be made relative to those objectives.

Data. The information on initial costs of buildings needed for life cycle cost analysis is the same as that described in Chapters 2 and 3 in discussions of initial cost estimating; the same data sources and methods can be used. This applies to investment costs/initial costs including planning, design, and engineering, as well as site purchase, construction, installation, financing costs, carrying charges, and so on.

Future costs or costs-in-use were discussed in Chapter 5, including costs of operation, maintenance, repair costs, or functional use costs, property taxes, insurance, income taxes, and so on. For the estimate of future costs, inflation and escalation assumptions are significant. On the most general level, across-the-board general inflation can be applied to all elements in the analysis equally. If this is the case, the analysis can be simplified by ignoring inflation (expressing future costs in constant dollars, and then using real instead of nominal discount rates for the present worth or EUAC conversions). However, in reality, the costs for different elements often rise at different rates. In the past, for example, energy costs have at times risen at higher rates than other costs. If future costs have been estimated in current dollars (i.e., including inflation), a different inflation rate may have to be used for each analysis element, as applicable, and the resulting future costs then are discounted with nominal discount rates. If the second approach has been adopted, of dealing with future costs in terms of constant dollars (excluding inflation), a differential escalation rate must be applied to those elements whose inflation rates differ from general inflation. The differential escalation rate is the difference between general inflation and the expected inflation rate for the element in question. For example, if the general inflation rate is expected to be 6% but energy costs are expected to rise by 8% annually, the differential escalation rate applied to energy costs should be 2%.

Replacement Periods. Here we are concerned with assumptions regarding the frequency with which various subsystems in a building must be replaced. The question of the trade-off between a higher initial cost for better-quality materials or specific subsystems, for example, and their longer replacement periods, often represents the most direct connection between architects' actual design decisions and life cycle cost considerations, even where no overall LCCA is being performed for the project.

Resale/Salvage Value. The salvage or resale value of an item plays a larger role in life cycle cost analysis involving machinery or vehicles, for example, than for buildings. Production machinery made of metal always can be sold at least for scrap at the end of its useful life, perhaps even for second-hand use; so its salvage value must be estimated and included in the analysis. Buildings or building parts that have to be demolished or replaced at the end of their useful life (for their current owner) usually do not command much in the way of resale value. There are exceptions, of course: a building that may have to be replaced and sold because it can no longer serve as the principal place of production or operation for a specific business, but could be reused for other purposes, or a historic building with precious "antique" architectural detail. But these circumstances are not often encountered in the practice of life cycle analysis for new projects. Even where there is material that could be used as scrap material (e.g., steel) the costs of demolition and its removal from the site often exceed any expected salvage value. Thus, life cycle cost analysis for building projects will be concerned primarily with the costs of disposal: dismantling the building and hauling away the debris.

If there is a significant resale value, as might be the case with the site itself, so that capital gain would be realized for that element of the analysis, capital gains taxes must be considered, in addition to sales tax. The question arises of whether federal income tax should be included. If the project is a commercial building from which the owner derives, say, rental income, then income tax can be seen as part of the costs of conducting that business. But this tax is related to income, not directly to a single identifiable element of the life cycle cost analysis; if income drops or rises, the tax will also. These changes could be the result of decisions and developments quite unrelated to the project itself.

If it is necessary to include consideration of the tax implications of the project in the analysis, its character will shift from a life cycle cost analysis to a more general economic performance analysis using different measures of performance, such as a rate of return or benefit–cost ratio.

Outcome: Measure of Performance

Life Cycle Cost Analysis involves calculation of the Life Cycle Cost (LCC) either as Present Worth or EUAC for all alternatives to be compared. The decision rule is to choose the solution with the lowest LCC. Whenever the comparison is between a baseline alternative (the conventional solution, e.g., for a HVAC system) and new energy-conserving alternatives, interest will focus on the amount the alternative system is expected to save over the life span and its relation to the added investment needed for the nonconventional solution: the Savings–Investment Ratio (SIR), which was defined in Chapter 7 as the ratio of the difference between the Life Cycle Costs of two solutions over the difference in initial costs or initial investment required. If the SIR is to be established, the LCC part of the analysis must be done by using the Present Worth measure. Here, the decision rule is that the SIR must be at least 1.00, or, more precisely, one plus the investor's discount rate, in order for the alternative solution to be worthwhile. Among a set of alternatives, the solution with the highest SIR should be selected. However, it may be appropriate to perform an incremental analysis on the results of a SIR analysis.

The Life Cycle Cost Analysis Process

The following is a brief summary of the steps involved in carrying out a complete life cycle cost analysis:

1. Establishment and description of the objectives to be served by the analysis, its scope, the alternatives to be considered, and the constraints for the decisions.
2. Selection of the measure to be used: Present Worth or EUAC; SIR (as applicable).
3. Establishment of the basic assumptions for the

analysis (see the above discussion), and of the breakdown of system components to be included in the analysis;

4. Compilation of data:
   - Size, units of alternatives (quantitative description of all alternatives for analysis).
   - Initial cost data.
   - Future cost data and time of occurrence of expected costs; replacement periods.
   - Salvage costs and/or values.

   It is a good habit at this point to establish a cash flow diagram for each alternative.
5. Discounting of all cash flows item by item to convert amounts to PW or EUAC for all alternatives.
6. Calculation of LCC for all alternatives.
7. Calculation of Savings–Investment Ratio (SIR) for all alternatives, as applicable.
8. Comparison of the LCC and the SIR for all alternatives; review and evaluation.
9. If necessary, examination of results for risk and uncertainty aspects, and establishment of relevant decision-theoretic criteria in preparation for the decision (see Appendix 11).
10. Review and decision. (Are the results acceptable? If not, can the solutions be modified for better results? Repeat the analysis as needed.) Selection of the preferred solution.

Figure 8-9 shows an example of a life cycle cost analysis.

RATE OF RETURN ANALYSIS

Concept

As a technique of performance analysis, the examination of the rate of return generated by an income-producing building has gained importance with the advent of and improvement of computer equipment. Return rates are better measures of performance than, for example, payback, and lend themselves to year-by-year analysis in any desired degree of detail. As explained in Chapter 7, a number of rates of return can be used; their common denominator is that they all conceptually measure the project's return (net income, cash flow, after-tax income, cumulative present worth returns plus reversion, etc.) over the equity contribution invested, or over the total capital investment (Total

Study period = expected life is: 40 years

COMPONENTS>>	SITE Land	Landscaping	Parking & drives	Lighting landscap. & (site area) parking	Signage	BUILDING Foundation	Superstruct.	Ext. walls: windows	Ext. walls: opaque	Roof	Partitions	Finishes	Elevators	Plumbing	HVAC	Lighting/ electrical	Fixed equipm.	
Unit	squ. ft.	squ. ft.	squ. ft.	squ. ft.	squ. ft. site	squ. ft.	squ. ft.	squ.ft.	squ.ft.	squ.ft.	squ.ft	squ. ft. TFA #		squ.ft	squ. ft	squ.ft.	squ. ft	
Area/amount	100000	5589.28	84000	89589.28	100000	10410.72	104107.2	13040	39120	10410.72	43580	104107.2	3	104107.2	104107.2	104107.2	104107.2	
(Formula or source)	SITAR	LSAR	PARKAR	LSAR+ PARKAR	SITAR	FNDAR	TFA	EWARW	EWARQ	RFAR	IWALLAR	TFA	ELEVNO	TFA	TFA	TFA	TFA	
Year of acquisition	0	3	2	3	2	2	2	2	3	2	2	3	2	2	2	3	3	
component life: % of st. per.	100%	50%	50%	25%	25%	100%	100%	50%	100%	25%	50%	25%	50%	50%	25%	25%	25%	
Initial cost yr. 0:																		
Unit price	$5.00	$5.00	$4.25	$0.10	$0.05	$5.00	$12.00	$20.00	$16.00	$7.00	$2.00	$6.00	$294,000	$4.00	$10.50	$5.40	$1.80	
Amount	$500,000	$27,946	$357,000	$8,959	$5,000	$52,054	$1,249,286	$260,800	$625,920	$72,875	$87,160	$624,643	$882,000	$416,429	$1,093,126	$562,179	$187,393	
differential escalation rate	0	0	0.01	0.01	0	0	0	0	0	0	0	0	0	0.01	0.02	0.1	0	
Amount at acquis.	$500,000	$27,946	$364,176	$9,230	$5,000	$52,054	$1,249,286	$260,800	$625,920	$72,875	$87,160	$624,643	$882,000	$424,799	$1,137,288	$748,260	$187,393	
PW Equiv. year 0 (base year)	$500,000	$22,185	$312,222	$7,327	$4,287	$44,628	$1,071,062	$207,031	$536,626	$62,479	$69,190	$495,862	$756,173	$364,197	$975,041	$593,993	$148,759	Subtotal PW
No. replacements (incl. first acqui)	1	2	2	4	4	1	1	2	1	4	2	4	2	2	4	4	4	
Total replacement PW Initial Cost	$500,000	$44,370	$624,444	$29,310	$17,147	$44,628	$1,071,062	$414,063	$536,626	$249,914	$138,381	$1,983,448	$1,512,346	$728,393	$3,900,164	$2,375,972	$595,034	$14,765,300
Component life (yrs)	40	20	20	10	10	40	40	20	40	10	20	10	20	20	10	10	10	
EUAC for period																		
1	$41,930	$1,860	$26,183	$614	$359	$3,742	$89,819	$17,362	$45,002	$5,239	$5,802	$41,583	$63,413	$30,542	$81,767	$49,812	$12,475	
2	$0	$399	$6,854	$314	$167	$0	$0	$3,725	$0	$2,427	$1,245	$19,261	$13,605	$7,995	$46,168	$59,845	$5,778	
3	$0	$0	$0	$161	$77	$0	$0	$0	$0	$1,124	$0	$8,922	$0	$0	$26,068	$71,898	$2,676	
4	$0	$0	$0	$82	$36	$0	$0	$0	$0	$521	$0	$4,132	$0	$0	$14,719	$86,378	$1,240	
5	$0	$0	$0	$0	$0	$0	$0	$0	$0	$0	$0	$0	$0	$0	$0	$0	$0	Subtotal
Subtotal EUAC for Init. cost	$41,930	$2,260	$33,037	$1,172	$639	$3,742	$89,819	$21,087	$45,002	$9,311	$7,047	$73,898	$77,018	$38,537	$168,722	$267,933	$22,169	$903,324

Annual costs:	SITE Land	Landscaping	Parking & drives	Lighting landscap. & (site area) parking	Signage	BUILDING Foundation	Superstruct.	Ext. walls: windows	Ext. walls: opaque	Roof	Partitions	Finishes	Elevators	Plumbing	HVAC	Lighting/ electrical	Fixed equipm.	
Maintenance: Unit price, annual	$0.00	$0.50	$0.10	$0.20	$0.05	$0.00	$0.01	$0.60	$0.05	$0.25	$0.08	$0.50	$200.00	$0.30	$0.50	$0.80	$0.75	
Maintenance: annual cost	$0.00	$2,795	$8,400	$17,918	$5,000	$0	$1,041	$7,824	$1,956	$2,603	$3,486	$52,054	$600	$31,232	$52,054	$83,286	$78,080	
Repair & replacement unit price	$0.00	$0.30	$0.50	$0.15	$0.05	$0.00	$0.10	$0.50	$0.05	$0.85	$0.05	$0.15	$1,000.00	$0.60	$0.75	$1.25	$0.40	
Repair & Replacement: cost	$100,000	$5,590	$84,001	$89,589	$100,000	$10,411	$104,107	$13,041	$39,120	$10,412	$43,580	$104,107	$1,003	$104,108	$104,108	$104,108	$104,108	
Energy/fuel type 1: cost / unit	$0.00	$1.50	$0.20	$0.00	$0.00	$0.00	$0.50	$1.35	$0.05	$0.35	$0.00	$0.00	$0.00	$0.01	$1.00	$0.00	$0.10	
Energy cost type 1	$0.00	$8,384	$16,800	$0	$0	$0	$52,054	$17,604	$1,956	$3,644	$0	$0	$0	$1,041	$104,107	$0	$10,411	
Energy / fuel type 2: unit price	$0.00	$0.01	$0.01	$0.00	$0.25	$0.00	$0.01	$0.00	$0.00	$0.00	$0.00	$0.20	$250.00	$0.20	$0.50	$2.50	$0.60	
Energy cost, type 2	$0.00	$56	$840	$0	$25,000	$0	$1,041	$0	$0	$0	$0	$20,821	$750	$20,821	$52,054	$260,268	$62,464	Subtotal
Subtotal annual	$100,000	$16,824	$110,041	$107,507	$130,000	$10,411	$158,243	$38,469	$43,032	$16,658	$47,066	$176,982	$2,353	$157,202	$312,322	$447,662	$255,063	$2,129,836

Total EUAC	$3,033,160
PW annual costs	$36,169,262
PW Initial Costs	$14,765,300
Total PW LCC	$50,934,562

FIGURE 8-9. Life cycle cost analysis (for office building shown in previous examples).

Initial Project Cost)—the rate at which the money invested returns to the investor.

There are two forms of analysis of this kind—a shortcut version and an in-depth, year-by-year examination of the initial and future cash flows. The former establishes an "average" annual rate of return based on the revenue and cost picture of a typical year of operation after the project has reached the stabilized expected level of occupancy and a steady balance of revenues and expenses. This kind of analysis would not consider reversion in the return picture, that is, the proceeds from sale of the project.

The second approach calculates the rate of return on a year-by-year basis with actual current dollar revenues and expenses and can do this both for the ongoing operation and for the assumption that the project is sold at the end of the respective period. The procedures for the two kinds of analysis will be slightly different.

Assumptions

The assumptions that must be made and clarified in preparation for a rate of return analysis are basically the same as those that go into initial cost estimates, cost-in-use estimates and life cycle cost analysis, and estimates of the revenues for the project. Again, it is important to clarify the purpose of the analysis: what are the owner's concerns and objectives with regard to the building? These factors determine the specific rate chosen as the measure of performance that serves as the basis for the required decision. In the example shown below in Figure 8-10, a number of return rates are calculated in one spreadsheet, to provide a better overview. They are not all needed for one to meaningfully compare alternative solutions and make decisions.

Of course, all needed costs and revenues must be established. For future amounts, assumptions must be made about the rates of inflation and escalation and the manner in which inflation should be treated—whether costs should be expressed as current or constant dollars; and the baseline year must be specified as appropriate. The discount rate for conversion of future amounts into EUAC or present worth equivalents must be determined.

Process for Rate-of-Return Analysis

A. Average Annual Return for Typical Year

The steps involved in carrying out a calculation of rate of return for the typical stabilized year of operation are the following:

1. Clarify the decision problem, objectives, and purpose of the analysis. Select an appropriate rate as the performance measure.
2. List and describe the alternatives to be compared.

3. Establish the relevant assumptions for the analysis. Set the threshold value of the return rate at or below which any solution should be rejected. Develop revenue and expense cash flows for a sufficient number of years to be sure that a stable level of operation has been reached.
4. For all alternatives that are compared, determine the cash equity contribution or the total project cost as appropriate to the chosen measure of performance.
5. For all alternatives under scrutiny, calculate the average annual return for the chosen measure in constant dollars for the year in which the equity contribution is made. Form the ratio constituting the rate chosen as the performance measure.
6. Compare the rates for the alternative solutions. The solution with the highest return rate is the preferred candidate for implementation. In the case of one solution only, determine whether the return rate is lower or higher than the established cutoff rate. Discuss the results, and decide (on a solution, or whether to adopt or reject a solution.)

B. Steps for Year-by-Year Analysis of Return Rates

1. Clarify the decision problem, objectives, and purpose of the analysis. Select an appropriate rate as the performance measure.
2. List and describe the alternatives to be compared.
3. Establish the relevant assumptions for the analysis. Set the threshold value of the return rate at or below which any solution should be rejected. Determine the study period, planning horizon (i.e., the period of crucial concern for the investor), inflation and escalation rates, depreciation and appreciation rates, and discount rate. Develop revenue, expense, and debt service cash flows for a sufficient number of "startup" years to ensure that a stable level of operation has been reached before projecting the further cash flows with a stable pattern.
4. For all alternatives that are compared, determine the cash equity contribution or the total project cost as appropriate to the chosen measure of performance.
5. For all alternatives under scrutiny, having estimated initial project cost and equity, establish a complete cash flow picture (it is a good habit to draw a cash flow diagram) for the entire study period or planning horizon.
6. Calculate the annual return for the chosen measure in current dollars for the year in which cash flow occurs.
7. Convert all future cash flows into present value of the baseline year in which the equity contribution is made. Form the ratios constituting the rate chosen as the performance measure, year by year.

	Year i	1	2	3	4	5	6	7	8	22	23	24	25
POTREVi	Gross revenues (in 1000's) = POTREV(1)*(1+INFL)^i	1,560.00	1,752.82	1,857.98	1,969.46	2,087.63	2,212.89	2,345.66	2,486.40	11.52	5,958.81	6,316.34	6,695.32
VLOSS(i)	Vacancy loss (in 1000's) = POTREV(i)*VR	124.80	140.23	140.23	140.23	140.23	140.23	140.23	140.23	140.23	140.23	140.23	140.23
ANREV(i)	Annual effective revenues = POTREV(i) - VLOSS(i)	1,435.20	1,612.59	1,717.76	1,829.24	1,947.41	2,072.66	2,205.44	2,346.18	5,481.29	5,818.58	6,176.11	6,555.09
ANEXP(i)	Annual expenses (in 1000's) = ANEXP(1)*(1+INLF)^i	438.39	492.58	522.13	553.46	586.67	621.87	659.18	698	1,579.77	1,674.55	1,775.03	1,881.53
NOI(i)	Net Operating Income (1000's) = ANREV(i) - ANEXP(i)	996.81	1,120.01	1,195.63	1,275.78	1,360.74	1,450.79	1,546.26	1,64	3,901.53	4,144.03	4,401.09	4,673.56
DS	Debt service (i) (in 1000's)	732.57	732.57	732.57	732.57	732.57	732.57	732.57	0	0.00	0.00	0.00	0.00
CTO(i)	Cash Throw-off(i) (1000's) = NOI(i) - DS	264.24	387.45	463.06	543.21	628.17	718.23	813.69	.75	3,901.53	4,144.03	4,401.09	4,673.56
SVAL(i)	Site value (i) (1000's) = SCST *(1+SAPPR)^i	500	551.25	578.81	607.75	638.14	670.05	703.55	72.98	1,462.63	1,535.76	1,612.55	1,693.18
BVAL(i)	Building value (i) (1000's) = (TIPRJCST-SCST)*(S-i)/S	6,504.82	6,215.71	6,071.16	5,926.61	5,782.06	5,637.51	5,492.9	.469.24	3,324.68	3,180.13	3,035.58	2,891.03
SALVAL(i)	Project sale value(i) (1000's) = BVAL(i)+SVAL(i)	7004.82	6766.96	6649.98	6534.36	6420.20	6307.56	6196.		4862.22	4787.31	4715.89	4648.13	4584.21
SALCST(i)	Sale cost(i) (1000's) = SALVAL(i) * 0.06	420.29	406.02	399.00	392.06	385.21	378.45	37	291.73	287.24	282.95	278.89	275.05
MORTBAL(i)	Mortgage balance (in 1000's) DS*((1+MLINT)^(MP-i)-1)/(MLINT*(1+MLINT)^(MP-i))	5395.92	5310.86	5215.60	5108.91	4989.41	4855.57	4	0.00	0.00	0.00	0.00	0.00
NETSALVAL(i)	Net Sale Value (i) (1000's = SALVAL(i) - MORTBAL(i)-SALCST(i)	1,189	1,050	1,035	1,033	1,046	1,074	4	4,570	4,500	4,433	4,369	4,309
PWANREV(i)	PW Effective income = ANREV(i) / (1+DISC)^i	1328.89	1382.54	1363.61	1344.55	1325.37	1306.13	.33	1025.68	1008.23	990.99	973.97	957.16
PWANCST(i)	PW of Costs year i (1000's) =(ANEXP(i)+DS)/(1+DISC)^i	1084.22	1050.36	996.02	945.27	897.85	853.5	58.82	296.07	290.58	285.20	279.92	274.74
CUMPWANREV(i)	Cum. PW benefits up to yr i =SUM(PWANREVi)	1328.89	2711.43	4075.04	5419.58	6744.96	8051	339.98	25365.66	26373.89	27364.89	28338.85	29296.02
CUMPWBEN(i) if sold year i	Cum. PW benefits(i) = CUMPWANREV(i)+(NETSALVAL/(1+DISC)^i)	2272.83	3611.70	4896.95	6179.16	7456.56	877	25336.35	26273.61	27201.64	28119.88	29027.88	29925.23
CUMPWANCST(i)	Cumulative PW Costs yr i = SUM(PWANCST(i))	1084.22	2134.59	3130.61	4075.88	4973.73	5	14415.38	14711.45	15002.03	15287.23	15567.15	15841.89
CUMPWCST(i) if sold year i	Cumul. PW Costs(i) =CEQC+CUMPWANCST(i)	2452.19	3502.55	4498.57	5443.84	6341.69		15783.35	16079.41	16370.00	16655.20	16935.12	17209.86
COCR(I)	Cash On Cash rate of return = CTO(i) / CEQC	0.19	0.28	0.34	0.40	0.46	4	1.99	2.68	2.85	3.03	3.22	3.42
CUMCTO(i)	Cumulative CTO up to yr i = SUM (CTO(i))	264.24	651.69	1114.75	1657.96	2286.	.44	26226.80	29899.54	33801.07	37945.10	42346.19	47019.75
BTEP(i)	Before Tax Equity Payback yr i =SUM(CTO(i))/CEQC	0.19	0.48	0.81	1.21		17.18	19.17	21.86	24.71	27.74	30.96	34.37
PWCTO(i)	Present worth CTO year i = CTO(i) /(1+DISC)^i	244.67	332.17	367.59	399.28	4	584.09	584.50	729.61	717.65	705.79	694.05	682.42
CUMPWCTO(i)	Cumulative PW CTO yr i = SUM(PWCTO(i))	244.67	576.84	944.43	1343.71		9340.09	9924.60	10654.21	11371.86	12077.65	12771.70	13454.12
DISCBTEP	Discounted BTEP = SUMPWCTO(i) / CEQC	0.18	0.42	0.69	0.98		6.83	7.26	7.79	8.31	8.83	9.34	9.84
PWNETSALVAL	Present worth Net sale value = NETSALVAL(i)/((1+Disc)^i)	1100.56	900.28	821.92	759.5		942.24	996.36	907.95	827.75	755.00	689.03	629.21
SUMPWBEN(i)	Sum of all PW Returns to yr i = PWNETSALVAL + CUMCTO(i)	1345.23	1477.12	1766.35	2103.	3	10282.33	10920.96	11562.16	12199.61	12832.65	13460.73	14083.34
CEQC	Equity investment	-1367.97	-1367.97	-1367.97	-136	.97	-1367.97	-1367.97	-1367.97	-1367.97	-1367.97	-1367.97	-1367.97
	CUMPWCTO(i)	244.67	576.84	944.43	134	.01	9340.09	9924.60	10654.21	11371.86	12077.65	12771.70	13454.12
	PWNETSALVAL(i)	1100.56	900.28	821.92	7	.33	942.24	996.36	907.95	827.75	755.00	689.03	629.21
IRR	Internal Rate of Return yr i for cash flow only	-82.11%	-57.83%	-30.96%		.08%	582.77%	625.50%	678.84%	731.30%	782.89%	833.63%	883.51%
	IRR cash flow plus reversion	-0.92%	4.90%	19.37%		3.10%	592.72%	635.40%	687.27%	738.51%	789.10%	838.99%	888.17%
		1	2	3		18	19	20	21	22	23	24	25
BCR(i)	Benefit-Cost Ratio year i	0.93	1.03	1.09		1.56	1.58	1.61	1.63	1.66	1.69	1.71	1.74

FIGURE 8-10A. Rates of return for an office building project—solution A. Partial spreadsheet with return rates highlighted.

	Year i	1	2	3	4	5	6	7	8	22	23	24	25
POTREVi	Gross revenues (in 1000's) = POTREV(1)*(1+INFL)^i	1,560.00	1,752.82	1,857.98	1,969.46	2,087.63	2,212.89	2,345.66	2,486.40	41.52	5,958.81	6,316.34	6,695.32
VLOSS(i)	Vacancy loss (in 1000's) = POTREV(i)*VR	124.80	140.23	140.23	140.23	140.23	140.23	140.23	140.23	140.23	140.23	140.23	140.23
ANREV(i)	Annual effective revenues = POTREV(i) - VLOSS(i)	1,435.20	1,612.59	1,717.76	1,829.24	1,947.41	2,072.66	2,205.44	2,346.18	5,481.29	5,818.58	6,176.11	6,555.09
ANEXP(i)	Annual expenses (in 1000's) = ANEXP(1)*(1+INLF)^i	507.19	569.88	604.07	640.31	678.73	719.46	762.62	808.?	1,827.67	1,937.33	2,053.57	2,176.78
NOI(i)	Net Operating Income (1000's) = ANREV(i) - ANEXP(i)	928.01	1,042.71	1,113.69	1,188.93	1,268.68	1,353.21	1,442.82	1,537	3,653.62	3,881.26	4,122.54	4,378.31
DS	Debt service (i) (in 1000's)	843.08	843.08	843.08	843.08	843.08	843.08	843.08	? 0	0.00	0.00	0.00	0.00
CTO(i)	Cash Throw-off(i) (1000's) = NOI(i) - DS	84.93	199.63	270.61	345.84	425.59	510.13	599.73	1.88	3,653.62	3,881.26	4,122.54	4,378.31
SVAL(i)	Site value (i) (1000's) = SCST *(1+SAPPR)^i	500	551.25	578.81	607.75	638.14	670.05	703.55	92.98	1,462.63	1,535.76	1,612.55	1,693.18
BVAL(i)	Building value (i) (1000's) = (TIPRJCST-SCST)*(S-i)/S	7,564.96	7,228.74	7,060.63	6,892.52	6,724.41	6,556.30	6,388.19	,034.65	3,866.54	3,698.43	3,530.32	3,362.21
SALVAL(i)	Project sale value(i) (1000's) = BVAL(i)+SVAL(i)	8064.96	7779.99	7639.44	7500.27	7362.55	7226.35	7091.?	5427.63	5329.17	5234.19	5142.87	5055.38
SALCST(i)	Sale cost(i) (1000's) = SALVAL(i) * 0.06	483.90	466.80	458.37	450.02	441.75	433.58	42?	325.66	319.75	314.05	308.57	303.32
MORTBAL(i)	Mortgage balance (in 1000's) DS*((1+MLINT)^(MP-i)-1)/(MLINT*(1+MLINT)^(MP-i))	6209.95	6112.07	6002.43	5879.64	5742.12	5588.09	5? ?	0.00	0.00	0.00	0.00	0.00
NETSALVAL(i)	Net Sale Value (i) (1000's = SALVAL(i) - MORTBAL(i)-SALCST(i)	1,371	1,201	1,179	1,171	1,179	1,205	?98	5,102	5,009	4,920	4,834	4,752
PWANREV(i)	PW Effective income = ANREV(i) / (1+DISC)^i	1328.89	1382.54	1363.61	1344.55	1325.37	1306.13	43.33	1025.68	1008.23	990.99	973.97	957.16
PWANCST(i)	PW of Costs year i (1000's) =(ANEXP(i)+DS)/(1+DISC)^i	1250.25	1211.38	1148.79	1090.34	1035.72	984.66	529.87	342.53	336.18	329.96	323.85	317.85
CUMPWANREV(i)	Cum. PW benefits up to yr i =SUM(PWANREVi)	1328.89	2711.43	4075.04	5419.58	6744.96	8051.?	24339.98	25365.66	26373.89	27364.89	28338.85	29296.02
CUMPWBEN(i) if sold year i	Cum. PW benefits(i) = CUMPWANREV(i)+(NETSALVAL/(1+DISC)^i)	2621.36	3741.20	5010.68	6280.02	7547.15	881?	25455.13	26379.20	27295.33	28202.86	29101.22	29989.90
CUMPWANCST(i)	Cumulative PW Costs yr i = SUM(PWANCST(i))	1250.25	2461.63	3610.43	4700.77	5736.49	67?	16633.89	16976.41	17312.60	17642.55	17966.40	18284.25
CUMPWCST(i) if sold year i	Cumul. PW Costs(i) =CEQC+CUMPWANCST(i)	2824.59	4035.97	5184.77	6275.11	7310.83	? ?6	18208.23	18550.75	18886.94	19216.89	19540.74	19858.59
COCR(i)	Cash On Cash rate of return = CTO(i) / CEQC	0.05	0.13	0.17	0.22	0.27	.40	1.52	2.18	2.32	2.47	2.62	2.78
CUMCTO(i)	Cumulative CTO up to yr i = SUM (CTO(i))	84.93	284.56	555.17	901.02	1326.61	?4.94	21338.14	24777.02	28430.64	32311.90	36434.44	40812.75
BTEP(i)	Before Tax Equity Payback yr i =SUM(CTO(i))/CEQC	0.05	0.18	0.35	0.57	0.?	12.03	13.55	15.74	18.06	20.52	23.14	25.92
PWCTO(i)	Present worth CTO year i = CTO(i) /((1+DISC)^i)	78.64	171.15	214.82	254.21	28?	510.25	513.46	683.15	672.05	661.04	650.12	639.31
CUMPWCTO(i)	Cumulative PW CTO yr i = SUM(PWCTO(i))	78.64	249.79	464.61	718.82	1?	7192.64	7706.09	8389.25	9061.30	9722.33	10372.45	11011.77
DISCBTEP	Discounted BTEP = SUMPWCTO(i) / CEQC	0.05	0.16	0.30	0.46		4.57	4.89	5.33	5.76	6.18	6.59	6.99
PWNETSALVAL	Present worth Net sale value = NETSALVAL(i)/((1+Disc)^i)	1269.55	1029.77	935.65	860.44	?	1052.79	1115.14	1013.54	921.43	837.97	762.36	693.89
SUMPWBEN(i)	Sum of all PW Returns to yr i = PWNETSALVAL + CUMCTO(i)	1348.19	1279.56	1400.26	1579.2?	60?	8245.43	8821.24	9402.78	9982.73	10560.31	11134.82	11705.65
CEQC	Equity investment	-1574.34	-1574.34	-1574.34	-1574.?	4.34	-1574.34	-1574.34	-1574.34	-1574.34	-1574.34	-1574.34	-1574.34
	CUMPWCTO(i)	78.64	249.79	464.61	718.?	?2.39	7192.64	7706.09	8389.25	9061.30	9722.33	10372.45	11011.77
	PWNETSALVAL(i)	1269.55	1029.77	935.65	860?	94.21	1052.79	1115.14	1013.54	921.43	837.97	762.36	693.89
IRR	Internal Rate of Return yr i for cash flow only	-95.00%	-84.13%	-70.49%	-54.?	4.46%	356.87%	389.48%	432.87%	475.56%	517.55%	558.85%	599.45%
	IRR cash flow plus reversion	-7.67%	-10.80%	-6.75%	?	38.85%	371.06%	403.55%	444.69%	485.56%	526.05%	566.11%	605.70%
		1	2	3		18	19	20	21	22	23	24	25
BCR(i)	Benefit-Cost Ratio year i	0.93	0.93	0.97		1.36	1.38	1.40	1.42	1.45	1.47	1.49	1.51

FIGURE 8-10B. Rates of return for an office building project—solution B. Partial spreadsheet with return rates highlighted.

8. Compare the rates and the trend patterns of the rates over the years for the alternative solutions. Are the return rate curves parallel, or do they intersect before the planning horizon? Draw a graph of the results for ease of comparison. The alternative achieving the highest rate of return consistently, or at the end of the predetermined planning horizon, is the preferred candidate for implementation. In the case of one solution only, determine whether the return rate is lower or higher than the established cutoff rate. Discuss the results, and decide.

Figures 8-10A and 8-10B show a spreadsheet model for the performance of two solutions of the familiar office building used in previous examples, with a selection (highlighted) of several kinds of return rates calculated year by year for a 25-year period. Note that the mortgage period is assumed here to be only 20 years, so that the debt service is zero after year 20. The spreadsheet (Figure 8-10A) shows the rates of return for a five-story solution with 20 modules per floor. The Internal Rate of Return (IRR) for cash flows and reversion (net sale value of the project at the end of the year i for which the rate is calculated) is negative in the first year, then quickly rises. It reaches 19.4% in the third year and 888% at 25 years. The corresponding BCR values for years 1, 3, and 25 are 0.93, 1.09, and 1.74, respectively.

An alternative solution with only ten modules per floor, and correspondingly ten floors instead of five (Figure 8-10B, but with all other assumptions being the same as above), shows a much less convincing picture. The IRR is negative through year 3 or longer; it reaches an acceptable level of 23% only in year 5, but then increases and reaches 606% in year 25. The Benefit–Cost Ratio shown for comparison at the bottom of the sheet is less than 1.00 for several years, and will be only 1.51 at year 25. Clearly, therefore, the first solution above is the better one.

VALUE ENGINEERING

Concept

In architecture, terms from other fields and disciplines often are used without much regard for their precise definition in the original field. Among these terms is "value engineering." In its loose sense, it refers merely to any attempt at reducing cost, but it can be much more than that. This section introduces the concept, its underlying principles, and possibilities for its application to architectural design: the concept of value and function as used in value engineering, including use functions, aesthetic functions, basic and secondary functions, and the notion of worth, as well as value analysis, the value index, and the Functions Analysis Systems Technique (FAST).

What does value engineering mean? It is true that it aims at reducing cost. In a more general sense, it deals with manipulating—improving—the relationship between the value and the cost of a product. This in itself could be seen as just another way of describing the design activity and the concern of any responsible designer. However, as a technique (some call it a recognized discipline in its own right, with a national organization, SAVE—Society of American Value Engineering) that is taught and applied in many engineering fields, it means something much more specific: value engineering is a systematic effort to assist the design process in the improvement of a product under development, using a specific methodology, conceptual framework, and corresponding set of techniques to increase value by looking for less expensive means for carrying out a required function, and to do this in such a way that the results can be measured.

Value engineering goes back to World War II efforts to ensure the production of needed parts of engines and electric devices in the face of shortages of critical materials. Its main aim at that time was to identify the function of various machine parts, for example, and systematically to look for substitute materials and design solutions that could perform the same function at less cost and using less expensive materials.

Conceptual Framework

Value engineering approaches its goals by first identifying, distinguishing, naming, and analyzing the functions of the product, system, or building to be designed: function analysis. There are two kinds of functions, use functions and aesthetic functions. Use functions are defined as involving some action that must be performed, expressed by a verb and its object. For example, the function of a column can be said to "support vertical load." Aesthetic functions are, broadly speaking, those functions that please or otherwise gratify the client, owner, users, and others. They too are expressed by means of a verb and a noun. An example might be to "emphasize the importance of the entrance." Functions then may be further classified as basic or primary, that is, essential or unequivocally required by client or user, and as secondary, that is, merely desirable but not essential, or arising because of an element that was selected to perform a primary function. For example, because an air-conditioning system is chosen that generates noise, it becomes necessary to design its enclosure so as to shield the adjacent rooms from that noise.

Functions—both use and aesthetic functions—are associated with a cost. This is either the cost of the existing element that is used to provide a given function, or the standard estimate of the conventional way in which the function is being fulfilled. The cost is the present or initial cost for the component in question.

For a component serving more than one function, its cost can be prorated among the functions.

The way in which value is dealt with in value engineering cannot escape the problems discussed in the chapters on evaluation, measures of performance, and the issue of money as a yardstick of value. In value engineering, different types of value are recognized: use value (the main concern of value analysis in engineering), esteem value, cost value (the costs needed to produce a value), and exchange value (Riggs and West, *Engineering Economics*). The problem lies in finding a quantifiable measure of value. As a rule, when the noun in the expression of a function refers to something that has a measurable aspect to it, it is easy to develop a measure for the value. For example, the noun "load" in the above function of a column "support vertical load" represents a ratio scale variable with units of, for example, pounds. In the language of the evaluation approach discussed in Appendix 6, we would say that the load in this case provides a criterion that could be used to objectify a judgment made about the "goodness" of the way the building element satisfies the function.

Two other concepts are used in connection with the determination of value of a function: one is worth, defined as the lowest cost at which a function can be fulfilled in the most elementary manner with current means and technology (see Kirk and Spreckelmeyer, *Creative Design Decisions*). The other is the relationship or the ratio of the worth and the cost of a given function in a proposed design solution. In the literature, this ratio sometimes is worth/cost and sometimes cost/worth, the value index. If the ratio is 1 or close to it, the function has good value—it is provided at minimum cost. If it is greater than 1, the value is poor and calls for a value engineering effort.

Some graphic techniques used in connection with value engineering should be mentioned: for example, the graphic function analysis technique (Kirk and Spreckelmeyer), which plots the worth and cost of various subsystems (e.g., of a building) and therefore allows a visual assessment of the areas with greatest potential for further improvements (Figure 8-11) and the Function Analysis Systems Technique (FAST). The latter is a diagramming technique that depicts the logical relationships between the functions involved in a complex subsystem or component (Figure 8-12).

Application to Architectural Design

Although some authors such as Kirk and Spreckelmeyer advocate the use of Value Engineering (VE) and techniques such as FAST for architectural programming and early schematic design, it is probably the spirit and attitude behind VE that is most valuable for application in architecture. The definition of adequate

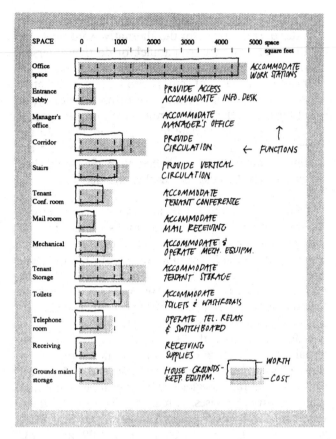

FIGURE 8-11.   Graphical function analysis (Kirk and Spreckelmeyer, *Creative Design Decisions*).

measures of performance (that would combine economic and other concerns) is in itself problematic, but it is necessary to find some resolution to that issue for all those projects where part of the performance is actual monetary return. Function and worth analysis would introduce a whole new and different set of concepts into the game, whose connection to the existing measures is easy to understand only for the cost part of the equation.

The attitude of consciously looking at the purpose or function of a building or building part and actively searching for ways of achieving the same purpose with cheaper or better solutions certainly should be part of every design effort. The new tools of quick analysis of design solutions by means of computer programs and spreadsheet models can help turn this attitude into action even when other measures of performance are used than those suggested by function analysis—measures that, as stated before, must be developed anyway in most projects.

As for the steps of actual application of value engineering, because it is not a single specific technique but involves an entire set of different tools, techniques, measures of performance, and procedures, the reader should consult the considerable literature on the subject.

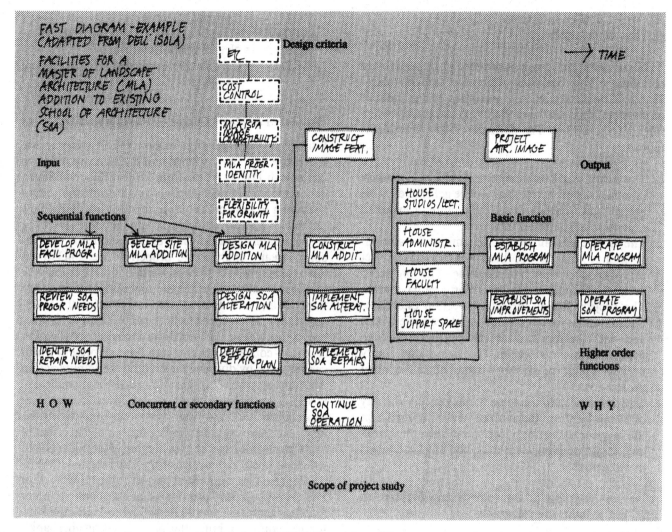

FIGURE 8-12.   Function analysis systems technique (FAST) diagram (after Dell'Isola and Kirk, *Life Cycle Costing for Design Professionals*).

THE REAL ESTATE "PRO FORMA" ANALYSIS AND STATEMENT

Overview

The results of the various forms of performance analysis discussed thus far, and even of financial feasibility analysis that will be presented in Chapter 9, may not be sufficient to guarantee that a project can be implemented successfully. For example, the performance analysis for a planned commercial building may indicate that its expected long-term economic performance —say, the projected rate of return—indeed is very worthwhile; but the owner may not have enough cash at hand for the required equity contribution, or the revenues from the project may not be enough to cover all expenses and income taxes for the owner during the early years of its operation.

The real estate pro-forma analysis is an attempt to look at all parts of the financial package needed to implement a building project, to assure the owner (and

lender) that all essential conditions for the viability of the project are given. The resulting pro forma statement is a standard requirement in real estate project proposals. It looks at the project from four main points of view, which are represented in the four main parts of the analysis. This section gives a brief overview of the concept and steps of the pro forma analysis. Most of the terms and variables used in the pro forma should be familiar to the reader from previous chapters: calculation of initial cost (budget), annual expenses, determination of equity, mortgage and debt service, estimation of income, cash flows, project value, and calculation of tax implications of the project.

Parts of the Pro Forma Analysis

There are four main parts of a pro forma analysis:

1. Income and expense
2. Value and loan

3. Project budget and equity
4. Tax implications

These parts are largely independent of each other although results from some parts are used in others. The level of detail can vary according to the amount of information available in the phase of a project when the analysis is done. The purposes of the parts are as follows:

1. The income and expense statement seeks to ascertain the level of income that will be generated by the project and to establish a clear picture of expected expenses in relation to income.
2. The value and loan section determines the amount of money that can be borrowed for the project (mortgage), its relation to the value of the project to an investor, and the resulting mortgage payments.
3. The project budget and equity section establishes the amount of money needed to get the project built and, with the result of the calculation of the loan from the previous section, the size of the equity needed, that is, the investor's own share in the initial project cost.
4. The tax implication statement analyzes the viability of the project from the point of view of its impact on the investor's taxes, or, more specifically, the spendable cash remaining after taxes, and any tax benefits it will generate.

The various parts of the analysis are discussed in detail in the following paragraphs.

Pro Forma Analysis Part I — Income and Expenses

The potential gross annual income is estimated, based on the project's Net Leasable Area and the assumed rental rates. The latter must be obtained from a market study. Losses from vacancy and bad debts are subtracted by using estimated vacancy rates, again based on market data. The annual expenses then are estimated, either as a percentage of the annual income or in a more detailed fashion by using data for operation, management, maintenance, repair/replacement, heating and cooling costs, and so forth, on a dollar per square foot of Total Floor Area basis. Subtracting expenses from the annual revenues, the result is the Net Operating Income. The following are the equations for this part:

$$NOI = POTREV*(1 - VR) - EXP \qquad (8\text{-}2)$$

$$POTREV = NLA*MRR \qquad (8\text{-}3)$$

where:

NOI	= Net Operating Income
VR	= Vacancy Rate (%)
EXP	= Annual Expenses
POTREV	= Potential Gross Annual Income ($)
NLA	= Net Leasable Area (sf)
MRR	= Competitive market rental rate ($/sf of NLA/yr)

Pro Forma Analysis Part II — Project Value and Loan

Using the Net Operating Income (NOI) from above, the project's value is determined. It is important to distinguish Project Value from Project Cost or Budget. A commercial project's value is what someone should be willing to pay for it considering the anticipated flow of rental income over the years, and therefore should be approximately equal to the present value of that income stream. If this were properly done, one would have to establish the present worth of the incomes, year by year, of all the years of the remaining useful life of the building, at a discount rate representing the investor's reasonable expectation of rate of return. In practice, a shortcut to this process of income capitalization is used — converting future cash flows to present worth capital value by dividing the NOI by the Capitalization Rate or cap rate (CAPR) (also called the overall rate, OAR). The cap rate represents the investor's earnings expectation and may vary from one investor to the next. Another way of expressing the same idea is that of multiplying NOI by the Net Income Multiplier (NIM), which is simply the inverse of the Cap rate or 1/CAPR.

The resulting capitalized project value (CAPVAL or PROJVAL) is the basis for what the bank or other lending institution should be willing to lend the investor to acquire or build the project. Using the bank's loan-to-value ratio, the loan amount or mortgage amount is established. Given the interest rate and mortgage term, we can find the annual mortgage payment or debt service, using the capital recovery factor or mortgage constant, as it is also called. Subtracting the debt service from the NOI gives the (before tax) cash flow or cash throw-off CTO. These are the equations for this part:

$$CAPVAL = NOI/CAPR \qquad (8\text{-}4)$$

$$LOAN = LVR*CAPVAL \qquad (8\text{-}5)$$

$$DS = LOAN*CRF$$
or:
$$= LOAN*(LINT*(1 + LINT)^m / ((1 + LINT)^m - 1)) \qquad (8\text{-}6)$$

$$CTO = NOI - DS \qquad (8\text{-}7)$$

where:

CAPVAL	= Capitalized Value
CAPR	= Capitalization Rate

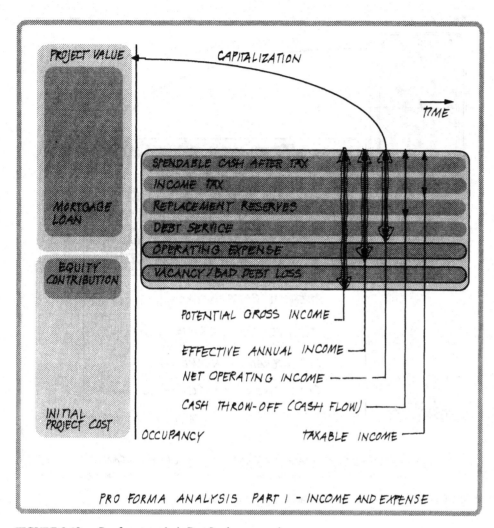

FIGURE 8-13.   Pro-forma analysis Part I—income and expenses.

LVR	= Loan-to-Value Ratio
CRF	= Capital Recovery Factor
LINT	= Loan Interest Rate
m	= Mortgage Term
DS	= Debt Service (Annual Mortgage Payment)
CTO	= Cash Throw-Off (Cash Flow)

Note that the practice of determining the loan amount based on the value of the project may be different from the way we often think about this—the loan being a percentage of the project cost. The latter reflects the bank's desire to make the owner assume a substantial equity stake in the project; the former is grounded in its concern for not lending more than what the project would bring in if it had to be sold (if the owner should default on the mortgage payments).

Pro Forma Analysis Part III—Budget and Equity

The project budget is an estimate of the amount of money needed to build the project, which is then set aside to cover the initial cost. In previous chapters, the task of estimating this—Building Cost, Project Cost, Total Initial Project Cost—was discussed in detail. No repetition of that discussion is needed here, except to point out that the level of detail in this section is likely to vary more than in the other parts, depending on the stage of the project and on the amount of equity required. The equity is established by subtracting the loan from the project value as determined in the previous element above:

$$CEQC = PROJCST - LOAN \qquad (8\text{-}8)$$

Pro Forma Analysis Part IV—Tax Implications

This section was much more important before the tax reforms of the 1980s; earlier tax laws often were designed to encourage investments in real estate, and thus the real advantages of a project often became visible only after scrutiny of the tax picture. The main attractive feature—tax credits for projects with negative cash flow—has largely been eliminated.

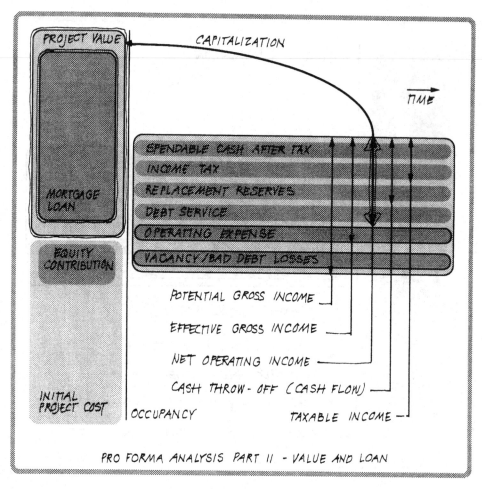

FIGURE 8-14.   Pro-forma analysis Part II—value and loan.

Calculation of the tax benefits arising from a project starts with the NOI. The tax laws up to 1986 allowed the investor to deduct as expenses not only the "depreciation loss" of the building but also the interest portion of the debt service paid every year. The 1986 revisions in the tax laws changed the terms for depreciation allowances for the building and also for deducting the interest on mortgage payments (which now is allowed only for home mortgages for one or two residences), resulting in less favorable conditions. Tax changes are likely to occur again in the future, and it is important either to follow these developments closely or to engage the services of a good tax specialist. The remaining benefits of building projects after taxes are the most important consideration for the investor, but they are not a prime focus of our attention here because they depend on so many factors not under the architect's control that the after-tax bottom line is not a good measure of performance of the architect's effort.

Depreciation can be calculated according to one of several different methods. The IRS rules currently allow use of the simplest, straight-line, method with 27.5 years for residential and 32.5 years for commercial construction.

Although depreciation arrived at by the straight-line method is the same year after year, the amount of interest paid each year as a part of the debt service payment varies and must be established separately for each year (last year's mortgage balance times the interest rate).

The tax rate depends on the owner's overall income level or tax bracket. The tax liability is calculated by multiplying the taxable income by the tax rate.

The tax liability or credit is subtracted from or added to, respectively, the Cash Throw-Off to find the Spendable Cash After Taxes or After Tax Cash Flow. This figure divided by the owner's equity investment is the yield, an important after-tax measure of performance. These are the corresponding equations:

$$\text{TAXINC} = \text{NOI} - \text{DEPREC} - \text{INTAMT} \quad (8\text{-}9)$$

$$\begin{aligned}\text{DEPREC} &= \text{BLDCST}/27.5 \text{ for residential construction}\\&= \text{BLDCST}/32.5 \text{ for commercial projects}\\&\quad \text{(simple straight line)}\end{aligned} \quad (8\text{-}10)$$

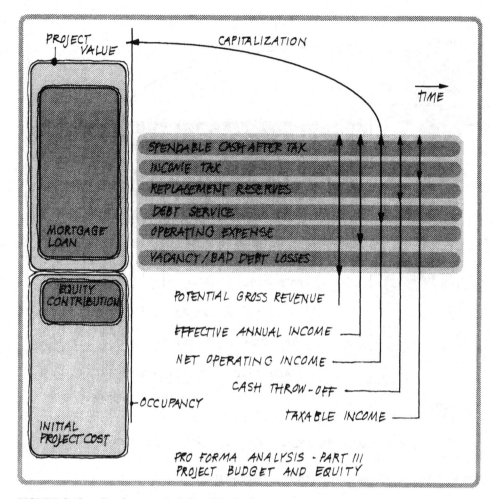

FIGURE 8-15.   Pro-forma analysis Part III—budget and equity.

$$INTAMT_i = MORTBAL_{i-1}*MINT \qquad (8\text{-}11)$$

$$TAX = TAXINC*TAXRT \qquad (8\text{-}12)$$

$$SCAT = CTO - TAX \qquad (8\text{-}13)$$

$$YIELD = SCAT/CEQC \qquad (8\text{-}14)$$

where:

TAXINC	= Taxable Income
DEPREC	= Depreciation Allowance
$INTAMT_i$	= Amount of interest paid in year i
BLDCST	= Building Initial Cost
$MORTBAL_i$	= Mortgage Balance for year i
TAXRT	= Tax rate (income tax)
SCAT	= Spendable Cash After Taxes
MINT	= Mortgage loan interest rate

It should be obvious that the pro forma analysis in its simple form as presented is a rather crude tool. It rests on the simplifying assumption of unchanging annual cash flows for both income and expenses. The variables involved in all those sections involving future income and expenses are likely to vary from year to year, especially during the first few years before the project will have reached its projected steady-state occupancy rate. There may be unusual expenses involved in fine-tuning various subsystems of the building, or straightening out problems remaining from construction. Therefore, to achieve a comprehensive picture, the pro forma analysis should be done not only for one year but at least for a number of years, up to a steady state of operation, and ideally for the entire planning horizon. If rents or expenses are subject to significant differential variations (e.g., different inflation rates to be applied to different cost elements) a continuous year-by-year analysis is advisable. Using computers, this is no longer a prohibitively difficult task.

STUDY QUESTIONS

1. Reexamine the table in Figure 8-1, and modify it in view of the three main types of task: decision, selection, and design. What difference does this modification make in the sequence of steps for each technique? In the measure of performance used?

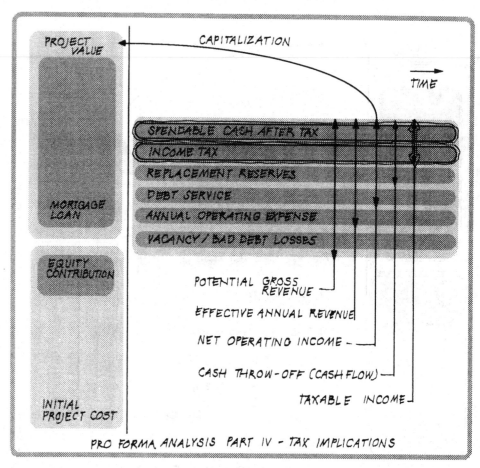

FIGURE 8-16.   Pro-forma analysis Part IV—tax implications.

2. In theory, a benefit–cost analysis resulting in a BCR of just over 1.00 is acceptable because the benefits are greater than the costs. Why should the threshold or cutoff value for acceptable solutions be somewhat higher, and by how much?

3. The aesthetic value of a building is an example of intangible benefits. Moreover, it is enjoyed to a large extent by people who may have no relationship to the project at all—people just passing by—and who also are not contributing to its cost (at least not directly in an easily traceable manner). Should the differing aesthetic values of several design alternatives be included in a Benefit–Cost Analysis for the project? If so, how could this be done? How would you determine the monetary value of this benefit?

4. Question 3 also illustrates the very common situation where costs and benefits of a building project are very unevenly distributed among the different parties or groups involved in the project or affected by it in some way. At the extreme, costs can be borne entirely by one group while benefits are enjoyed by others. How would you deal with this possibility in analyzing it (e.g., estimating magnitudes), possibly counteracting or compensating for it?

5. In the example for Benefit–Cost Analysis (Figure 8-2), the design variables for the architect are the same as those examined in earlier cost estimate models, namely, the dimensions of the building: the number of modules per floor, which indirectly determines the number of floors, and the length and the width of the module. If you generated the previous spreadsheets as you studied the initial and annual costs, it should not be too much trouble to add the necessary calculations for the Benefit–Cost Analysis, shown in Figure 8-2B. Make an educated guess as to how the solution presented in the spreadsheet should be changed to improve the BCR, and then implement the changes and see what happens.

6. Figure 8-4 demonstrates an example of incremental analysis. The solutions and assumptions are simplified; for example, no allowance is made for the fact that if covered parking is provided, a higher rental rate may be charged for the office space. But, then, the entire project may be aimed at a different type of tenant, so that the office space also must be built to higher standards and therefore higher cost. Introduce these considerations into the example, and ex-

plore what happens to the ordering of preferred solutions as a result.

7. The examples and graphs shown for breakeven analysis are quite simplified but serve to make an important point regarding the need to clarify client or investor objectives before any kind of analysis. Discuss the main aspects of client objectives that must be specified before the results of breakeven analysis can be used to make decisions between design alternatives.

8. Add the calculations highlighted in Figure 8-10 to your spreadsheet for the Benefit–Cost Analysis. Study the various return rates and what they tell you about the worthwhileness of solutions at different points in time. Then make educated guesses about design changes that might improve performance, make the respective changes, and test your hypotheses.

9
Financial Feasibility Analysis for Building Projects

THE CONCEPT OF FEASIBILITY

It may seem somewhat strange that the question of the feasibility of a building project was not taken up earlier; after all, if a project is not feasible, it does not do much good to calculate return rates and benefit–cost ratios for it. However, the definition of "feasibility" as used here, especially in connection with economic or financial feasibility analysis, is somewhat narrower than the word's usual sense. Few projects that we encounter in everyday practice will have features that make them entirely impossible (and therefore infeasible) to build. Thus, the question of feasibility is really about whether a project can be done without violating certain accepted rules, principles, or constraints. The constraints can range from logical consistency or physical possibility to relatively arbitrary criteria based on the goals and objectives of some involved party. In the case of economic feasibility or financial feasibility, at least some of the constraints involve measures of economic performance. This is the reason why those performance measures, and the corresponding analysis techniques, were discussed first.

Types of Feasibility

Before going on, it may be useful for us at least to take a look at some other types of feasibility, which now can be conveniently classified according to the kinds of constraints involved, to place the financial feasibility discussion in its appropriate larger context. As suggested above, in practice, few projects will have serious problems with logical or physical possibility feasibility, such as violating laws of logic or existence in space and time (for example, that two physical objects cannot exist in the same space at the same time); and only in very large or avant garde projects will we be likely to have to worry about technical, structural, or other building technology-related feasibility. By contrast, even small projects may run into issues of political and legal feasibility, with code and zoning regulation constraints being only the most visible obstacles that can prevent a project from going ahead. The proliferation of legislative, permit, and review requirements, for example, in the environmental or growth management domains has been decried by some as effectively making projects infeasible that were quite commonplace only a few years ago. For example, it is said that given today's more strict requirements for impact statements of various kinds, it would be so difficult and time-consuming to meet these demands that, for all practical purposes, the federal interstate highway system could not be built if it were started now.

Whereas codes and ordinances usually are well defined and present clear-cut limitations as to what is allowed and what is not, social or cultural feasibility problems are no less serious but often are very difficult to define and thus hard to deal with. In the latter situation there is a lack of clear thresholds in constraints, and such concerns often have a number of different aspects whose relationships are not well understood.

Economic or Financial Feasibility

This chapter discusses the somewhat narrow aspect of economic or financial feasibility, which, in the context of the preceding general remarks, can be seen to be defined by a concern for constraints that are measures of economic performance, and with the economic givens of the project delivery process—costs, levels of rental rates of competing projects, and market conditions. The concept and technique of financial feasibility analysis will be described as a particular way of assessing the economic viability of a project. The discussion assumes familiarity with the concepts of economic performance, using a performance benchmark as the starting point for the analysis, and with the basics of cost estimating and project financing, and will introduce additional concepts and variables as needed, such as income capitalization and estimated project value based on capitalized project income.

The discussion can conveniently be organized according to four main groups of analysis techniques, distinguished by the main outcome, as shown by the diagrams in Figures 9-1A through 9-1D. The measures of economic performance that were examined in Chapter 7 can be used to compare different projects to find out which is the more worthwhile, or to compare a proposed project with the returns from some standard average investment venture (e.g., savings or a stock market investment). They use assumptions about program, construction costs, and revenues (e.g., rental rates) to calculate expected performance. For example:

(a) Given: PROGRAM (or design solution),
               COST, and
               REVENUES
     Find: PERFORMANCE

Techniques aiming at an estimate or analysis of performance (understood as one of the performance measures discussed in Chapter 7 combining both cost and benefits) and starting with assumptions about cost and income, are represented by Figure 9-1A. They were discussed in Chapter 8.

There are other kinds of questions one might want to ask, representing the remaining arrangements of the four variables program, cost, income, and performance. For example:

(b) Given: PROGRAM (or design solution),
               COST, and
               PERFORMANCE
     Find: REVENUES (e.g., minimum rent that
               must be charged)

Figure 9-1B shows the logic of analysis efforts aimed primarily at an estimate of required income (e.g., rental rate), starting with assumed project cost. This is the underlying logic of "front door" financial feasibility analysis.

(c) Given: PROGRAM,
               REVENUES, and
               PERFORMANCE
     Find: COST (maximum acceptable construction budget)

Figure 9-1C deals with analysis starting with rental income or benefit assumptions and resulting in calculation of an affordable project cost or budget—"back door" feasibility analysis.

(d) Given: COST or BUDGET,
               INCOME, and
               PERFORMANCE
     Find: AFFORDABLE PROGRAM

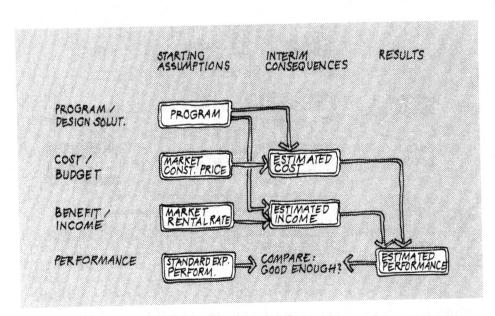

FIGURE 9-1A.   Analysis elements—performance analysis.

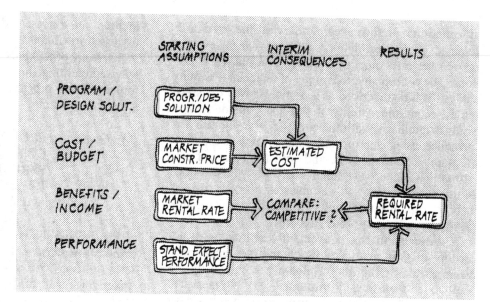

FIGURE 9-1B.   Analysis elements—"front door" feasibility analysis.

Figure 9-1D represents techniques using cost, income, and performance assumptions to arrive at an estimate of an affordable program.

Feasibility analysis is the kind of analysis aimed at answering these questions in general. However, financial feasibility analysis techniques in a more specific sense usually address only the second and third of these forms of analysis.

FEASIBILITY ANALYSIS: AUXILIARY CONCEPTS

Feasibility analysis involves some concepts and variables that we have not yet encountered. Foremost among them is the process of income capitalization— converting a stream of expected income in future years (net operating income) into the present value equivalent. This is done to find the value of a project (as opposed to its initial cost) for the purpose of establishing the amount that a lending institution should be willing to lend for its acquisition. To avoid confusion between value and cost, it helps to think of this concept in connection with the purchase of existing buildings. Because the construction cost was paid long ago, it is no longer relevant to establish the purchase price. The important question is: how much income can the property generate year by year over its remaining useful life?

Income capitalization refers to the process of con-

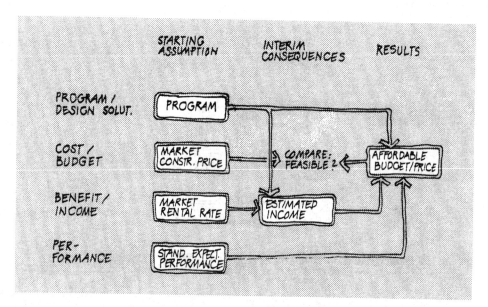

FIGURE 9-1C.   Analysis elements—"back door" feasibility analysis.

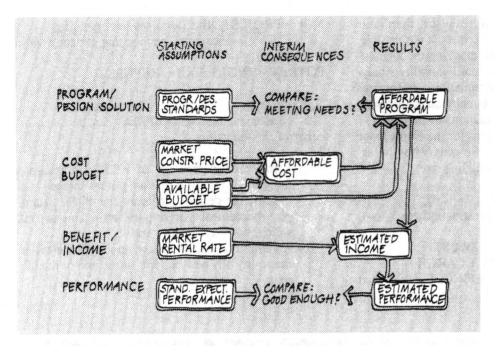

FIGURE 9-1D.   Analysis elements—affordable program analysis.

verting NOI payments into their present value equivalent. Project Value (PROJVAL) is the present worth equivalent of the stream of income payments over the years. More specifically, it is the present worth equivalent of the stream of Net Operating Income amounts (NOI) over the analysis period. The Capitalization Rate (CAPR) or cap rate is the rate at which the stream of NOI payments is converted into their present worth equivalent amount. It is estimated as the sum of the rate of return *on* an investment (discount rate) and the rate of return *of* the investment. The formula for establishing the cap rate from a known NOI and Project Value is:

$$CAPR = NOI/PROJVAL \qquad (9\text{-}1)$$

Sometimes the concept of Overall rate (OAR) is used interchangeably with that of the cap rate—the rate at which the capitalization process should be carried out, based on market analysis such as sales information.

$$OAR = NOI/Purchase\ price \qquad (9\text{-}2)$$

or, in the case of new construction:

$$OAR = NOI/TIPRJCST$$
$$\text{(total initial project cost)} \qquad (9\text{-}3)$$

Capitalization in perpetuity is the special case of a series of NOI's extending indefinitely without change into the future (i.e., over an infinite planning horizon). The formula for capitalization in perpetuity is:

$$PROJVAL = NOI/CAPR \qquad (9\text{-}4)$$

(The estimated project value is one year's NOI divided by the cap rate.)

The Net Income Multiplier (NIM) is the reciprocal of the overall rate, a ratio formed by the purchase price (or project value, assuming the two are equivalent) over the Net Operating Income:

$$NIM = Purchase\ price/NOI \qquad (9\text{-}5)$$

or, for new construction:

$$NIM = TIPRJCST/NOI \qquad (9\text{-}6)$$

The Building Budget (BLDBUD) is the amount of money available/required to construct the building, which should not be exceeded.

"FRONT DOOR" FINANCIAL FEASIBILITY ANALYSIS FOR BUILDING PROJECTS

The question of how much income is needed to pay for the project and to achieve the targeted economic performance is addressed by the Front Door financial feasibility analysis. It starts from the estimated initial cost for the given program or design solution, assuming an acceptable economic performance for the investment, and calculates the cost-justified rental income required to achieve this performance. The resulting rental rate is compared with rental rates prevailing in the market; if it is lower than rental rates charged by the competition, the project is deemed feasible; if it is higher than rents charged by the competition, the project is not feasible. Figure 9-2 shows a diagram summarizing the logic involved in carrying out a Front Door Feasibility Analysis. (Rectangular boxes stand for equations or processes that produce results; rounded bubbles stand for variable values produced by the equations or processes.)

The following are the equations for the Front Door Financial Feasibility Analysis. We start with the assumed construction price and building size:

$$BLDCST = BLDPR * TFA$$
$$\text{(required construction cost)} \quad (9\text{-}7)$$

$$PRJCST = BLDCST + SCST + SWCST$$
$$+ FEES + PRMCTS + CARCH$$
$$+ CONT + OHP + CFINCST \quad (9\text{-}8)$$

$$NOI = PROJCST/NIM \text{ (required net operating}$$
$$\text{income to pay for this project.)} \quad (9\text{-}9)$$

$$POTREV = (NOI + EXP + REPRES)/$$
$$(1 - VAC) \quad (9\text{-}10)$$

The replacement reserve (REPRES) can be taken as either of the following:

(a) Some fixed percentage of EXP (for example 0.05*EXP), spread out as an annual charge—an expense for periodic replacement of, for example, mechanical systems, the roof, and so on, which, however, is not spent every year.

(b) An annuity designed to save up the equivalent of the equity of a new building at the end of the life span, s.

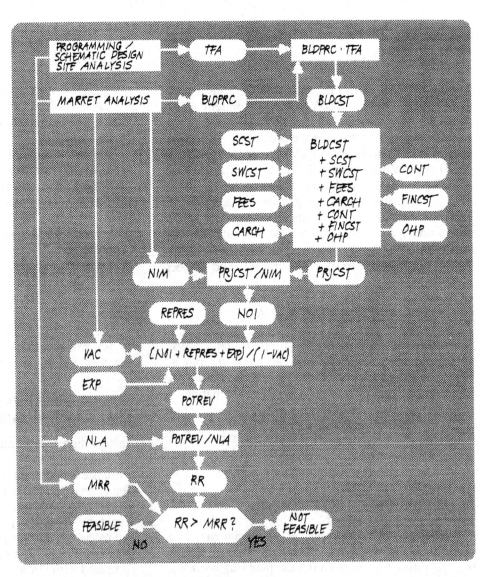

FIGURE 9-2. "Front door" financial feasibility analysis.

Then:

$$REPRES = CEQC*(1 + INFL)^{s}* $$
$$SINT/((1 + SINT)^{s} - 1) \quad (9\text{-}11)$$

$$RR = POTREV/NLA \quad (9\text{-}12)$$

RR is the required rental rate that must be charged to make the project feasible. If this rate is higher than the gong market rental rate (MRR), the project normally is not considered financially feasible.

The following are the steps involved in carrying out a front door financial feasibility analysis:

1. From the program or design solution, ascertain the Net Leasable Area (NLA) and the Total Floor Area (TFA).
2. From analysis of the market and available cost data for the give project type, establish:
   - The applicable building construction price, BLDPRC ($/sf of TFA).
   - The average competitive market rental rate, MRR ($/sf NLA/yr) for projects of similar size and quality in the neighborhood.
   - The expected average vacancy and bad debt rate, VAC (%), for this project type, quality, and neighborhood.
   - The expected economic performance for similar projects, as expressed by the Net Income Multiplier, NIM.
   - The expected rate of annual operating expenses ($/sf of TFA/yr) and therefore annual operating expenses, EXP, for projects of the given type.
   - Expected level of replacement reserves needed for the project.
3. Calculate the estimated initial building construction cost (BLDCST).
4. Calculate the expected total initial project cost (TIPRJCST) by adding the site cost, site work cost, fees, contingency allowance, carrying charges, overhead and profit, and financing costs to the building construction cost.
5. Divide TIPRJCST by NIM to obtain the annual required Net Operating Income (NOI).
6. Calculate annual expenses and vacancy losses, and add them to NOI to obtain Potential Gross revenues (POTREV).
7. Divide POTREV by the Net Leasable Area (NLA) to obtain the required annual Rental Rate, RR ($/sf/yr).
8. Compare the RR to the MRR. Interpret the results: is the project feasible or not feasible?

"BACK DOOR" FINANCIAL FEASIBILITY ANALYSIS FOR BUILDING PROJECTS

The Back Door Financial Feasibility Analysis reverses the logic of the Front Door Approach, starting from assumed competitive market rental rates (or purchase prices) and expected economic performance. It then calculates the maximum budget affordable cost that can be expended on the project for the given program or design solution. The resulting affordable construction price, in dollars per square foot of floor area, is compared with the prevailing construction prices on the market. If it is higher than market construction prices, the project is considered feasible (the income would allow a higher price); if it is lower, the project is not feasible: the market-oriented income is not sufficient to justify the construction cost one must expect to pay. Figure 9-3 shows the diagram for the Back Door approach to financial feasibility analysis.

The Back Door Financial Feasibility Analysis equations are as follows. Starting with an assumed competitive rental rate:

$$POTREV = RR*NLA \quad (9\text{-}13)$$

This is the potential gross revenue that the project could generate, before vacancy losses.

$$NOI = (POTREV*(1 - VACR)) $$
$$- EXP - REPRES \quad (9\text{-}14)$$

(See above for possible assumptions about REPRES.)

$$PROJVAL = NOI*NIM \quad (9\text{-}15)$$

This is the capitalized value of the project, that is, the present worth equivalent of the stream of NOI's over time, assuming a discount rate (expected earnings rate) of 1/NIM = cap rate; it is the amount one should be willing to pay to obtain those NOI's from the project. Note that this is a very different concept from PROJCST, the amount it takes to build or purchase a project.

$$BLDBUD = PROJVAL - SCST - SWCST - FEES $$
$$- CARCH - FINCST - OHP \quad (9\text{-}16)$$

This is the building budget; it is the maximum amount one can afford to pay for construction at the expected level of rental earnings.

$$TFA = NLA/NGR \quad (9\text{-}17)$$

Here The Net Leasable Area (NLA) divided by the standard expected Net to Gross rate (efficiency factor) (NGR) produces the required Total Floor Area.

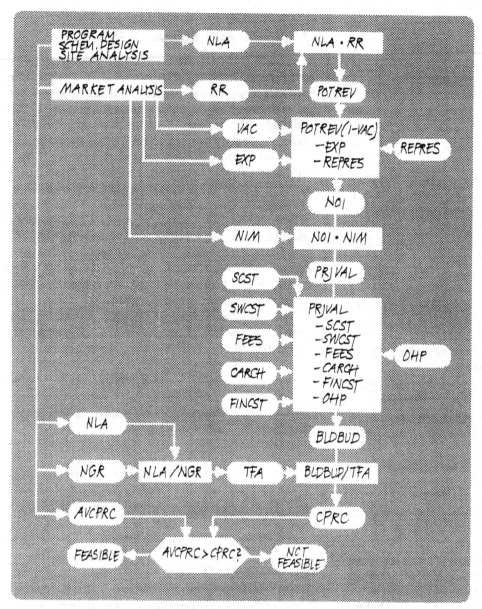

FIGURE 9-3. "Back door" financial feasibility analysis.

$$BLDPR = BLDBUD/TFA \qquad (9\text{-}18)$$

BLDPRC is the maximum income-justified construction price ($/sf of TFA) that one can afford to pay. If it is more than the average construction price for the type of building in question, the project is feasible; otherwise it is not.

Steps in Back Door Financial Feasibility Analysis

The following steps are necessary to carry out a back door feasibility analysis:

1. From the program or schematic design solution, ascertain the Net Leasable Area (NLA) and the Total Floor Area (TFA).

2. From an analysis of the market and available cost data for the give project type, establish:
   - The appropriate building construction price, BLDPRC ($/sf of TFA).
   - The average competitive market rental rate, MRR ($/sf NLA/yr) for projects of similar size and quality in the neighborhood.
   - The expected average vacancy and bad debt rate, VAC (%), for this project type, quality, and neighborhood.
   - The expected economic performance for similar projects as expressed by the Net Income Multiplier, NIM.
   - The expected rate of annual operating expenses ($/sf of TFA/yr) and therefore annual operating expenses, EXP, for projects of the given type.

- Expected level of replacement reserves needed for the project.
3. Calculate the estimated annual Potential Gross Revenues (POTREV).
4. Calculate and subtract the annual operating expenses (EXP) and replacement reserves (REPRES) to obtain the annual Net Operating Income (NOI).
5. Multiply NOI by NIM to obtain the capitalized project value (CAPVAL).
6. Subtract site cost, fees, and other project costs to obtain the net affordable building construction budget (BLDBUD).
7. Divide BLDBUD by the total floor area (TFA) to obtain the affordable construction price (CPRC).
8. Compare the affordable construction price (CPRC) with the prevailing market construction price (AVPRC) for comparable projects: is it higher or lower? Interpret the result and decide: is the project feasible or not feasible?

The front door and back door approaches can be combined, and also can be carried forward year by year. Whenever either income or annual expenses (or both) are expected to vary from year to year, it is advisable to carry out a time-projected analysis to make sure that the expected performance and thus the feasibility will be achieved throughout the study period.

CALCULATION OF YIELD WITHIN FINANCIAL FEASIBILITY ANALYSIS

To obtain the yield (after-tax cash flow over equity) from both front door and back door analysis, the following equations are used, where the letters F and B in the equation numbers denote that conceptually the same equation is used with different values in the front door approach (F) and back door analysis (B):

Front door:

$$FLOAN = LVR*PROJCST \qquad (9\text{-}19)$$

$$FCEQC = PROJCST - FLOAN \qquad (9\text{-}20)$$

$$FDS = MC*FLOAN \qquad (9\text{-}21)$$

Back door:

$$BLOAN = LVR*PROJVAL \qquad (9\text{-}19B)$$

$$BCEQC = PROJCST - BLOAN \qquad (9\text{-}20B)$$

$$BDS = MC*BLOAN \qquad (9\text{-}21B)$$

MC or CRF (capital recovery factor)
$$= LINT*(1 + LINT)^m/((1 + LINT)^m - 1)$$

$$INTAMT_1 = LOAN*LINT$$
(Amount of interest paid in year 1)   (9-22)

$$PRINRET_1 = DS - INTAMT_1$$
(principal retired in year 1)   (9-23)

$$FCTO_1 = FNOI_1 - FDS \qquad (9\text{-}24F)$$

$$BCTO_1 = BNOI_1 - BDS_1 \qquad (9\text{-}24B)$$

$$FTAXINC_1 = FCTO_1 + FPRINRET_1 + REPRES - DEPREC \qquad (9\text{-}25F)$$

and:

$$BTAXINC_1 = BCTO_1 - DEPR + REPRES + PPRET_1 \qquad (9\text{-}25B)$$

where:

$$DEPR = BLDCST/d$$

$$DEPREC = BLDBUD/d$$

and *d* is the depreciation period allowed by the tax laws.

$$FTXL = FTAXINC*TXRT \qquad (9\text{-}26F)$$

and:

$$BTXL = BTXINC*TXRT \qquad (9\text{-}26B)$$

$$FSCAT_1 = FCTO_1 - FTXL - CONT \qquad (9\text{-}27F)$$

and:

$$BSCAT_1 = BCTO - BTXL_1 - CONT \qquad (9\text{-}27B)$$

$$FYIELD_1 = FSCAT_1/CEQC \qquad (9\text{-}28F)$$

and:

$$BYIELD_1 = BSCAT_1/BCEQC \qquad (9\text{-}28B)$$

Figures 9-4 and 9-5 show the diagrams for yield, as derived from the front door and back door approaches, respectively.

EXAMPLE OF FEASIBILITY ANALYSIS

Figure 9-6 shows the result of Front Door and Back Door feasibility analysis applied to the familiar office building example. The design variable values and assumptions are the same as those used for the Rate of Return calculations in Chapter 8 (Figure 8-10).

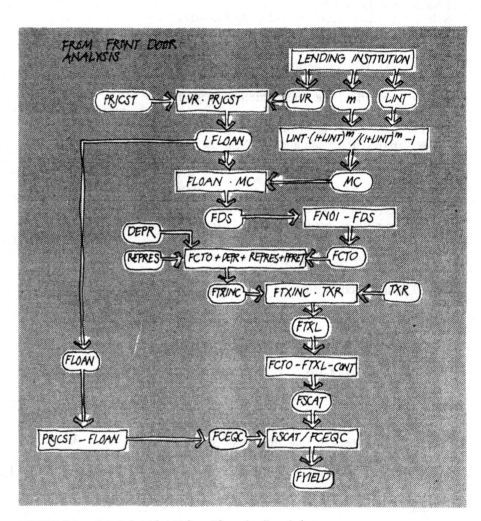

FIGURE 9-4.   Calculation of yield from "front door" analysis.

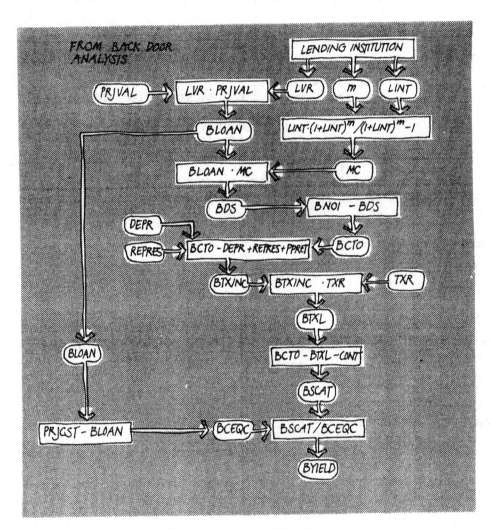

FIGURE 9-5. Calculation of yield from "back door" analysis.

FRONT DOOR FEASIBILITY ANALYSIS

Replacement reserves	REPRES	$68,398	CEQC/MP
Net Income Multiplier	NIM	9.00	
Required Net Operating Income	NOIR	$759,981	TIPRJCSTE/NIM
Required Gross Potential Rev.	POTREVR	$1,376,927	(NOIR+ANEXP+REPRES)/(1-VR)
Required rental rate	RRR	$22.95	POTREVR/NLA
Market rental rate	MRR	$25	
	FEASIBLE		=IF((MRR>RRR),"FEASIBLE","NOT FEASIBLE")

BACK DOOR FEASIBILITY ANALYSIS

Capitalized project value	PROJVAL	$8,971,258	NOI*NIM
Building budget	BLDBUD	$6,623,098	=PROJVAL-FINCSTE-SCST-SWCST-AEFEESE-LEGFEESE-ACCFEESE-UTCSTE-CCONTE-CARCHE-PERMCST-SVCST-LCCAFEE-SLTSTCST-FEQPCST-FURNCST-MOVCST-VARCST-DEVCST
Affordable constr. price	AFCPRC	$74.46	=BLDBUD/TFA
Market construction price	MCPRC	$53.00	
	FEASIBLE		=IF((AFCPRC>MCPRC),"FEASIBLE","NOT FEASIBLE")

FIGURE 9-6.   Feasibility analysis applied to office building example.

STUDY QUESTIONS

1. As an exercise, develop the steps needed for the analysis case shown in Figure 9-1D (affordable program analysis) in the same manner as those listed for the front door and back door analysis.
2. For both procedures, front door and back door approaches to feasibility analysis, an important further step should be added, whether the result was positive (i.e., the project was feasible) or negative: that of using the results of the analysis to look for ways in which the solution might be improved. If you developed the spreadsheet for the various analysis types following the discussion in preceding chapters including this one, try to manipulate the input variables for the example shown in Figure 9-6 to obtain better results.

10
Conclusion

REVIEW

Our text started out quite innocently, trying to explore some fundamentals of building economics. It first looked at the question of estimating as well as gaining some control over the initial cost of building projects, and then their future costs (or costs-in-use). Costs are justified by the benefits or value created with the project; so estimating these benefits or values was the next task. Trying to relate these benefits to the costs led to the concept of measures of performance, and considerable effort was devoted to study of a variety of such measures of economic performance of buildings, and to their use in a number of techniques and methods for the assessment and analysis of building project performance and worthwhileness.

This journey was guided initially by somewhat simple and straightforward practical motivations—one's desire to become, as an architect, a more competent partner in the discussions about economic considerations and concerns that influence the planning of building projects by clients and their financial advisors, and to gain better control over one's own efforts to meet those concerns, that is, to be able to design better.

It is now time to take stock of what was discovered along the way. Certainly, some answers were found that deserve to be summarized and evaluated; but also a number of new questions arose that are almost more significant and interesting than the answers. The itinerary was by no means as straightforward as the above brief outline and the sequence of chapters in the main body of the book suggest. Numerous detours into related topics and other disciplines or fields of study were necessary (mainly, the material provided in the appen-

dix). And the questions explored there—as well as a number of questions that could not be discussed but perhaps should be—make it clear that the subject of building economics is considerably more complex than it first appeared. In fact, our understanding of the very subject now appears in a different light: what is building economics?

This chapter takes a brief look at these issues.

WHAT WE FOUND: A SUMMARY OF ANSWERS AND RECOMMENDATIONS

With respect to the initial practical concerns, the reward of the journey is quite conspicuous—perhaps too much so. A first general observation is that architectural design decisions indeed have a considerable impact on the economic performance of buildings, but also that economic considerations and conditions, including decisions taken prior to design or outside the architect's sphere of influence, introduce constraints to architectural design. Then, however, it can be said that these mutual relationships are open to understanding and useful communication among those involved in planning a building. In spite of the initially confusing jargon and even the appearance of the mathematical formulae, tables of factors, and spreadsheets full of numbers, there are no really significant conceptual obstacles to understanding the fundamentals. The mathematics involved is on the whole not very high-level or demanding; at worst, it may be tedious. But the impressive array of concepts, tools, methods, and techniques that was assembled to help us with the tasks of estimating, evaluating, and controlling economic performance includes tools and methods that also address tedious tasks better than ever before—so much so that these tools (especially the computer) have the potential for radically changing the way we design. One of the main recommendations we take away from this part of the discussion is that estimating and analysis tasks that previously were so tedious, time-consuming, and thus costly as to prevent more than quite infrequent application now should become frequent, repeated companions to the planning, programming, and design process. It is fair to say that in the architectural practice of the future, cost estimates, feasibility analyses, and performance analyses will have to be done many times throughout the process, precisely because the data and tools (both hardware and software) will be so easily available that there will be no excuse for not using them. Thus, if the architect refuses to engage in the dialogue about how design decisions influence economic performance, she or he cannot be surprised if clients introduce constraints (regulations) aimed at guaranteeing the desired economic outcome but inadvertently and unnecessarily limiting the architect's design freedom.

The reader who may have expected more concrete practical design guidelines and tricks for making economically successful buildings may have been somewhat disappointed, but even that outcome can be seen as a positive result overall. Together with the tools, the perhaps newly acquired insight that practical results must grow anew in each new project from an open and informed discussion of assumptions, economic and other objectives, and appropriate means for achieving them, is an important one if it leads the architect and other participants to organize the programming and design process in such a way as to encourage and support that kind of constructive discourse.

All this can help make the architect who is willing to put forth the effort to master the jargon and the conceptual fundamentals a more responsible partner in the overall planning process of building projects, with potentially more influence over many areas that in recent times were given up to others. Incidentally, this possibility can open up and enlarge special new career opportunities and services within architecture. The question of whether the profession will or should eventually grasp these opportunities for additional in-house services, as opposed to leaving them to outside consultants, remains to be discussed and cannot be resolved here.

UNRESOLVED ISSUES AND NEW QUESTIONS

The availability and ease of application of the new estimating and analysis tools also gives rise to some very serious concerns. The answers mentioned in the preceding section—the techniques, analysis software,

measures of performance, and so forth, that can facilitate integration of economic performance analysis into the design process—are all answers at a comparatively low technological level. Therefore, their uncritical application in situations where there are more fundamental issues at stake can not only fail to adequately resolve problems, but actually can become misleading and make matters worse. Are there such fundamental issues in the set of topics surveyed? The answer is, unfortunately, yes. Especially the various side trips in the appendix, in retrospect and seen as a whole, suggest that not only are there some fairly wicked philosophical questions hidden underneath several of the concepts and techniques that seemed so helpful at first sight, but also that there are a number of additional issues that have not been identified, much less resolved, that often may have left the reader with a sense that the discussion somehow missed the essence of the problem.

Consider these issues that were not discussed but that one might reasonably expect a book such as this one to address: No mention was made of the housing shortage and affordable housing—when affordable housing is becoming so expensive that it is out of reach for a large segment of society; or of the growing problem of the homeless who have had to give up hope for any form of housing at all. Almost nothing was said about the environmental and ecological implications of building, about energy in construction and energy use in buildings, or about transportation and land use, all of which undeniably have economic implications. The deterioration of inner city areas and the psychological impact of urban and suburban developments are examples of other issues that could have been treated in this connection. Although good arguments can be made that a book such as this should say something about these issues, the choice here was to focus upon the concepts and tools needed to explore the problems, analyze the issues, and develop solutions.

The list of omitted items indicates that building economics is not the clean-cut subject it seemed to be at the outset of our journey, mainly concerned with the cost and performance of a single building. It overlaps with other fields in such a way that it is hard to draw clear borders. The economic impact of legislation, zoning, regulations and codes, the economic concerns of the workforce in the construction industry (labor economics), transportation economics, urban economics, the economics of the production of raw materials and building component manufacturing, and so forth, are all part of a larger set of issues that properly could be called building economics.

One might get a feeling for the interrelatedness of the various factors and parties involved in the overall economic process that brings buildings into being by exploring the network of contributors to a single building. Compare the simple diagram representing the almost

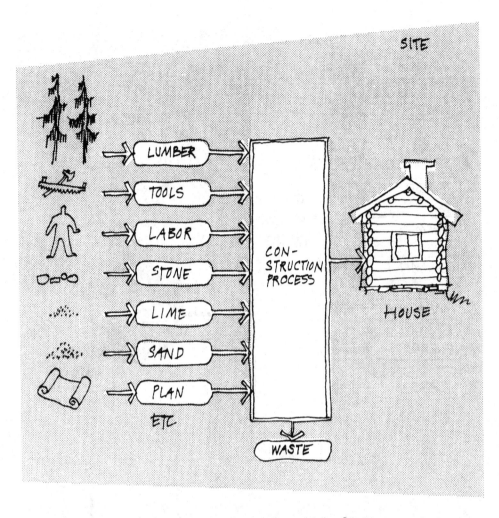

FIGURE 10-1.   A simple construction process involving no money flows.

mythical case of a pioneer building his or her own frontier home in the wilderness (Figure 10-1) with the production of a conventional suburban house (Figure 10-2). In the second diagram (but not the first) all service and material flows are accompanied by money flows in the opposite direction. Expanding the diagram as shown in Figure 10-3 to include the manufacturing of building components, materials, and so on, in one direction, and the processes of using, operating, maintaining, and repairing a building over time in the other, we may begin to get a sense of the complexity of the network or system that generates and sustains buildings.

Each contributor of service or materials must be compensated for his or her contributions, adding overhead, profit, taxes, and the cost of waste at each step. However, the exceptions to the general rule that all material flows are met by money flows in the opposite direction constitute interesting areas for possible improvement strategies. First, the basic raw materials are received "free." The money we pay for them is not going to "Mother Nature," but to the landowner, the workers, and the supplier of the equipment needed to

extract the material. Second, there is "sweat equity"— the work an owner and/or user will do without monetary compensation. Low-cost housing, especially in third world countries, would be unimaginable without this factor. Third, there is waste. It occurs as an output of virtually every transformation, in the form of wasted material, energy (heat dissipated), and time. The interesting aspect of this is that the substance wasted already has been paid for; then more money must be paid to dispose of the waste—money and material flow go in the same direction.

Still the overriding picture remains that of numerous steps, services, and tasks being performed by a large number of players, and the many associated money flows. Must we therefore conclude that we should work toward making that network smaller, eliminating steps, and "middlemen"? From the point of view of the individual client and his or her architect, such an attitude makes sense. Figure 10-4 shows a variation that might be called a strategy for systems planning to simplify the network, coordinate activities, and eliminate redundancies. But if one looks at the picture from say, a, government perspective, it is no longer so clear even

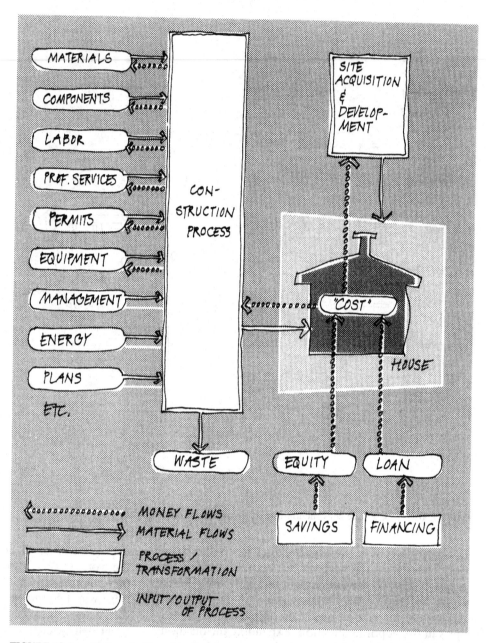

FIGURE 10-2. The production of a conventional house. Each material flow is accompanied by a flow of money in the opposite direction.

what is "cost" and what is "benefit"; if an industry supports so many people now, does it make sense to work for a situation where it supports fewer? (Of course, but only if there are other jobs for people to move into who will have lost their jobs.) The pitfalls of policymaking at these levels is illustrated by the story of a research effort in a European country to study the large-scale application of industrialized building systems, which was supported by government officials with the argument that the many small "inefficient" building contractors had to be killed off to make the industry as a whole more efficient and productive. The result of this was that after some years it was getting almost impossible to find people for small-scale building upkeep, repair, and maintenance—which was how the inefficient contractors survived winters when freezing weather prohibited new construction projects. These services were something the streamlined large construction companies were not equipped to provide; instead they laid off workers, burdening the public with unemployment compensation as well as higher expenses for lower levels of maintenance.

The lack of straightforward recipes for economic-performance-oriented design is not only due to the preferred focus on the tools for developing solutions; it is also a reflection of several fundamental difficulties in

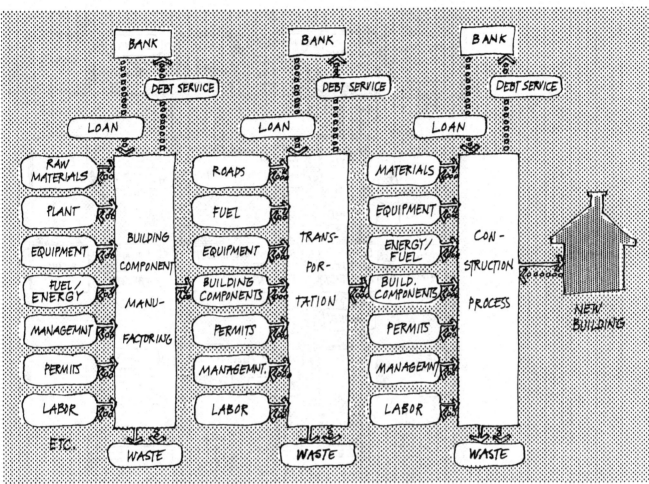

EMERGING NETWORK OF INPUTS AND MONEY FLOWS FOR BUILDING PROJECTS (SELECTIVE)

FIGURE 10-3A.   The building production system (simplified).

determining what qualifies as a good solution in any given situation, even where the problem is clearly within the domain of architectural decisions for a single project. Briefly these difficulties are related to the problem of evaluation, to the role of time and the difficulties of forecasting long-term performance, which in turn is related not only to the inherent uncertainties of the variables we are trying to predict but also to human activity and the urge to design — that is, to change the world as it presents itself to us, including our own and others' forecasts, plans, and designs.

The Problem of Value and Valuation

The main problem with economic evaluation lies in the relationship between general evaluation and values and the economic measures of performance. Although the problems of evaluation in general have been adequately recognized and discussed in theory and methodology, the task of translating the resulting insights into useful practical procedures still remains to be solved. It is fair to say that most attempts to introduce more thorough, systematic evaluation procedures into practical design and planning situations quickly run into the participants' impatience with the resulting cumbersome, tedious, and long-winded process. In such situations, one can see a growing urge to take recourse in methodological and conceptual shortcuts. If such shortcuts then offer themselves with the added lure of also appearing to be more "objective" and reality-oriented, they become quite irresistible. This often happens when general evaluation concerns compete for the participants' attention with "quantifiable" economic performance measures. The former in most cases cannot be measured with clear, objective, quantitative variables, and the relationship between these variables and the design variables that must be decided upon is difficult to model in the same straightforward way in which we have been able to model, for example, the life cycle cost of a building with a given shape. The cautions that are voiced, for example, in Appendix 3 about measurement, and especially the issue of money as a measure of

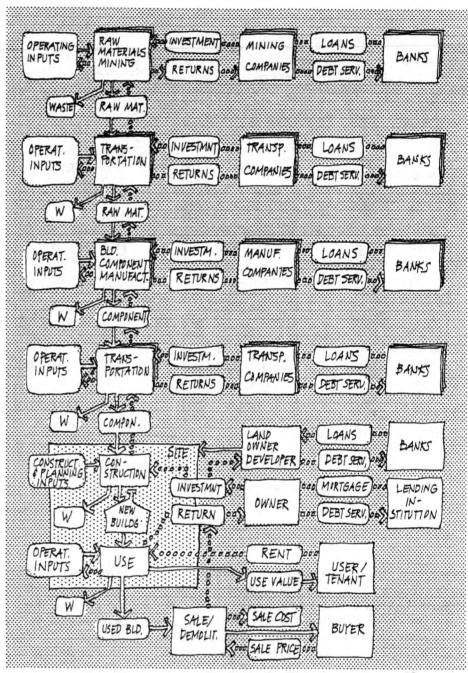

THE NETWORK OF FLOWS OF INPUTS AND MONEY FOR BUILDING PROJECTS

FIGURE 10-3B.   The network of input and output flows for construction of buildings—comprehensive.

value, then are easily forgotten, and the monetary, quantitative economic performance measures (which look objective even though they are not) are accepted as an adequate support for the decisions that must be made. They are given more weight in the decision outcome than they deserve. Moreover, economic valuations often tend to hide rather than reveal the true underlying values. For example, if a client organization

decides to hire an expensive well-known architect for the design of its new corporate headquarters, is this because of appreciation of that architect's design sensitivity and quality, or because the corporation wishes to hitch the architect's fame and name to the wagon of promoting its own image? The fees the client is willing to pay the designer say something about how much the organization values one service or the other, or some

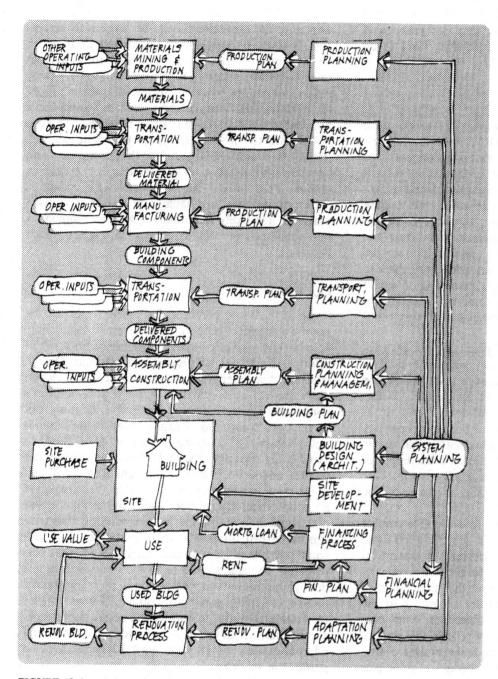

FIGURE 10-4.   A "systems planning" variation of the network.

combination of both, but not which service or which combination it is paying for.

These questions merely compound the more standard problems of evaluation that have to do with the fact that different parties involved in or affected by a project have quite different concerns, that the weights they assign to the respective evaluation aspects, even if they recognize each other's concerns, will vary considerably, and that quite often, the concerns are competitive in nature (aiming at win–lose outcomes among the

participants). They also serve to stress once again the issues surrounding optimization and attempts to optimize and to use claims of optimality to support particular solutions. As seen in Appendix 7, most optimization in practice tends to be suboptimization if only a limited set of variables is considered. If the focus already has been narrowed to easily measurable monetary performance variables (or performance measures derived from monetary measures), then only suboptimization is possible; and if the performance measures used only represent one group's concerns and point of view, such efforts can become quite misleading and even counterproductive.

THE ROLE OF TIME

The other major source of trouble has to do, once more, with time and our ability to forecast future developments. The discussion of forecasting in Appendix 10 clearly identifies the difficulties in forecasting even if the trends in question are objective phenomena subject to natural laws. When the variables to be predicted are the result of human activities (and, therefore, of human values, ambitions, etc.) the forecasting business becomes tricky indeed—in some respects akin to the problem of forecasting the way our own knowledge will grow in the future: if we could predict it, it would become today's knowledge, not tomorrow's, and therefore subject to further growth and change that we could not anticipate.

Add to this impossibility of predicting our own future knowledge the human habit, urge, and right to design—that is, to want to change and reshape the world in which we find ourselves. (One could call this one of the traits of the species that define us as humans.) Forecasting must assume that, somehow, our forecasts are exempt from becoming part of that world; otherwise, they will themselves become the target of efforts to change them. This is, of course, exactly what happens, to the detriment of the accuracy of the forecasts.

These difficulties have caused considerable trouble for social scientists—for example, they have led some to postulate an inherent "boundedness" of human rationality, or even irrationality in people to explain why people consistently fail to follow the predictions of what rational beings ought to do, according to the theories. Planning and design objectives, even economic ones, must be adopted by people; this implies that there must be a choice. The underpinning of that choice is not adherence to some theoretical principles of rationality, as much as economists try to find such principles and postulate these to be "intuitively acceptable" axioms, no longer to be questioned, upon which to base their theories. People's choices are at least as much driven by image—of who we want to be—as by rational pursuit of economic utility. And these images, for example, the ideal of individualism (which must find features of distinction in order to be realized, even if the distinctive features go against economic rationality), are adequate explanation for the contrariness and apparent irrationality of design choices, even economic choices. Postulating a set of principles for economic choice as rational, inevitable, and universal in itself makes it less likely that that set will be adopted by individuals as a personal basis for planning and design decisions.

OUTLOOK

Some of the above considerations seem to shroud the entire enterprise of economic planning and forecasting, of modeling and calculating performance, in a fog of futility. This is not an inevitable conclusion. On the contrary, the emerging story is one in which design plays the key role; and this is where the architect's contribution, as the expert on design and designing, can become much more significant and satisfactory than it now appears. The resulting recommendation is not to avoid using the concepts, tools, techniques, computer models, and so on, discussed throughout the book, but to use them in a different way. The goal is not to derive a solution in an algorithmic fashion by calculation and deduction from purported facts, a solution that then can be pushed through as the expert-sanctioned outcome over the objections and misgivings of other participants. The goal should be to use these tools to test our assumptions about what we think we know about a project, to explore possibilities, to use them as vehicles for communication, to explain views and preferences and the basis of judgment to one another in the overall effort to develop design solutions to which all involved parties can subscribe.

Floor Area Conventions and Building Efficiency

FLOOR AREA MEASUREMENT CONVENTIONS

The floor area is one of the most important concepts of cost and economic performance analysis for buildings. This is in part mere tradition; we are used to referring to the size of buildings in terms of their floor area. As such it is arbitrary—other countries use building volume rather than floor area for the same purpose. However, having been widely accepted as a convention it is now most convenient because many related matters are dealt with in the same way, ranging from regulations to construction cost data; all are expressed in terms of floor area or variables related to floor area. In the early stages of planning for a building—feasibility analysis, programming, schematic design—not much more than the approximate building size is known. Therefore, most of the forms of cost estimating and performance analysis at these stages use floor area as the conceptual unit of reference. It is important to understand the meaning of the various terms related to this concept and how they are measured and used: Total Floor Area or Gross Floor Area, Net Leasable Area, Net Assignable area, and Net Usable Floor Area, as well as the Floor Area Ratio and the Net-to-Gross Ratio or Efficiency Ratio.

Various institutions such as the American Institute of Architects (AIA) and the Building Owners and Managers Association (BOMA) have issued standards for measuring the various types of floor areas distinguished in buildings. The purpose of such standards is to ensure that calculations concerning the allowable building mass on a site or the areas of the building actually occupied and used by tenants (for which rent is paid) are made in a consistent and comparable manner.

Some of these standards differ from each other, and also with respect to the application of the terms to different building types. The important tasks for designers are to make sure that the project will be equal or better than the average expected efficiency level for the building type in question and to select the most efficient from a set of alternative schemes. The first task must use standard definitions; for the second, the consistent application of one definition to all comparable schemes is the critical issue.

Total Floor Area, Gross Floor Area, Architectural Area

The terms Total Floor Area (TFA), gross floor area, and architectural area refer to a measure of the building size in terms of floor area, as measured from (permanent) outside finished surface to outside finished surface of the building. It includes all areas occupied by walls, columns, stairs, ducts, and shafts, whether they can be physically occupied by people or not. AIA document D101 specifies how different parts of buildings should be counted in the TFA or architectural area for the purpose of estimating overall size and, from that, expected cost. It distinguishes regular floor space from parts of the building that are not fully enclosed, or are not complete floors:

- Typical floors, mechanical spaces, basements and subbasements, balconies in inside spaces, bulkheads and penthouse areas, enclosed entrances, duct spaces, and interstitial spaces higher than 6 feet count in full.
- Open terraces, pipe, and interstitial spaces less than 6

feet high and canopy areas are counted as 50% of their actual measured area.

- Open plazas, open roofs, parapets, voids in floor plans due to higher spaces, and foundations count 0%.

Net Usable Area, Net Leasable Area, or Net Rentable Area; Net Assignable Area

The area than a tenant can actually use, the net usable area (NUA) or net assignable area (NASF), is measured from the inside finish of external walls to the inside (tenant side) of permanent inside walls such as fire-rated corridor walls or party walls, walls of fire stairs or elevator shafts, and the like. However, structural elements such as columns or structurally necessary pilasters inside this area and areas of nonpermanent, nonstructural partitions (such as the tenant may install) are not subtracted. The net usable area does not include shared (public) corridors, restrooms, storage areas, janitors' closets, and so on. In principle, when several tenants share an office floor, the usable area is equal to the rentable or leasable area (NLA). If a single tenant occupies an entire floor, the elevator lobby, restrooms, closets, and so on, will be for the tenant's exclusive use; in this case the rentable area is different from the usable area and includes these areas. Often, however, the rentable area will include a prorated percentage of common corridors, rest rooms, lobbies, and so on, even where several tenants share these spaces.

Efficiency Ratio or Net-to-Gross Ratio

The Net-to-Gross Ratio (NGR) is formed by The Net Leasable Area over the Total (gross) Floor Area: NLA/TFA. Occasionally, the reciprocal is used for the same purpose: GNR = TFA/NLA. This book uses NGR or NLA/TFA exclusively. The higher the NGR is (that is, the closer to 1.00), the better, or the more efficient, the building.

USE OF FLOOR-AREA-BASED INDICATORS AND MEASURES

The Total Floor Area (TFA) is the basis of ratios and calculations of permissible urban density; for example, the Floor Area Ratio (FAR), where FAR = TFA/site area.

The net-to-gross ratio (NGR) is the most important indicator of floor plan efficiency and therefore also of expected initial cost. At all design stages, designers monitor the NGR achieved with the current scheme, trying to keep it within or better than standard levels of expectation for the building type. Everything else being equal, of two schemes with the same NLA, the one with

the lower NGR will have the lower TFA and therefore the lower initial cost.

Building efficiency as measured by the Net-to-Gross Ratio or its inverse, the Gross-to-Net Ratio, refers to the amount of space that must be built in the building as a whole (TFA) in order to provide the amount of space (net leasable area or net usable area) for which the owner will receive rental income, or which actually produces the use value of the building. The difference between the TFA and the NLA is the "nonrentable" area or service area. The cost of the building, and the relationship between cost and rental income, will be favorable if the NGR is high (close to 1); it will be the more unfavorable the more the NGR drops below 1. Because this relationship is almost always quite direct and close, the NGR can be used as a reliable performance indicator during the design process. The designer will not have to calculate the initial cost for each new design concept; it is sufficient to check the NGR to see if it is better or worse than preceding solutions. If the NGR does not improve, it is not worthwhile to further pursue the scheme.

Standard expectation levels of the NGR are further used at the feasibility analysis and programming stage to estimate the TFA, given the specified program requirement of NLA: TFA = NLA/NGR.

Standard Efficiency Expectations for Different Building Types

Based on experience with many examples of various building types, standard expected levels for the NGR have been established that can guide the programmer and the designer in many decisions during early development stages. For example, office buildings usually are expected to achieve an NGR of 0.75 to 0.80 to be considered sufficiently efficient; apartments from 0.67 to 0.80; hotels at least 0.62 to 0.70; schools 0.55 to 0.70, depending on the type; hospitals 0.55 to 0.67. These standards must be used judiciously. For example, the NGR for an office building that features mainly many small offices for single occupants will be lower than that for a building with large open office landscape type spaces, as the latter can be much deeper than the small offices. Using the same standard NGR expectation for both would be quite inappropriate.

DESIGN DECISIONS INFLUENCING EFFICIENCY

There are a few (almost obvious) design rules of thumb that can be applied to building efficiency and design decisions. One such rule is that because circulation spaces (hallways, corridors) are responsible for a significant fraction of nonrentable space in buildings, the spaces served by corridors should be as deep as possible compared with the corridor. Double-loaded corridors are more efficient than single-loaded corridors. If the spaces are specified to be rectangular, their short side should front to the corridor, as seen in Figure A1-1.

Another, perhaps less obvious, rule is that if the service core—elevators, fire escape stairs, duct space, and so on—of a building and its total required net leasable area are of a determined size, increasing the number of floors will reduce the NGR rather than improve it. This runs counter to an apparently widespread but largely fallacious view that high rise means high density and high efficiency. Multiplying the number of floors on which stairs, toilets, duct space, elevator lobbies, and so forth, must be provided, while allocat-

ing less net area per floor, clearly does not make a solution more efficient.

A fallacy sometimes encountered in connection with the effort to make a design scheme more efficient is based on the correct observation that efficiency is related to the ratio of the depth of served spaces—say, a hotel room—to the width of the corridor serving them. For example, in the process of modifying solution A in Figure A1-3, an alternative proposal may be solution B. Checking the NGR, it might look as though B is the better scheme because its NGR is higher. However, the higher NGR has been achieved not by a reduction in circulation space but by increasing the overall TFA. The rooms merely were made deeper—thus improving

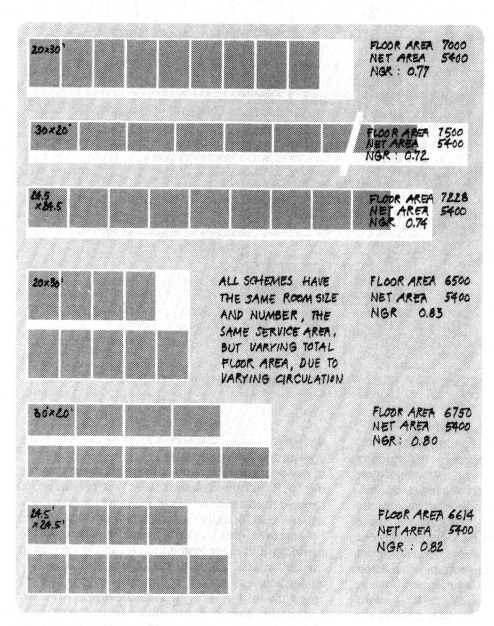

Figure A1-1A Effect of circulation and room proportion on NGR - A

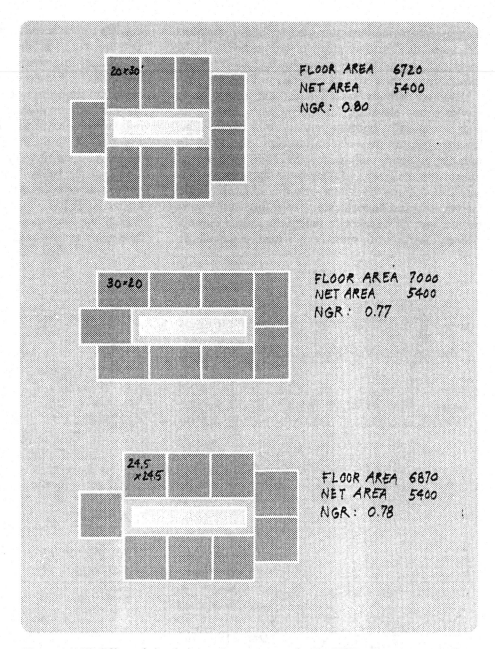

Figure A1-1B Effect of circulation and room proportion on NGR - B

the room depth to corridor width ratio—but not noticeably narrower, as there are constraints in furnishing hotel rooms preventing that. Thus the unintended results was that the overall floor area was increased, raising the total cost. As a consequence of making the rooms deeper, the unproductive circulation area inside the rooms was increased; the rooms lost appeal instead of gaining value. This kind of fallacious reasoning is not always as easy to spot as it is in this example. The lesson is to avoid uncritical reliance on a single performance indicator or measure, always double-checking results from different points of view. In this case, checking the allowable area per room for each scheme, together with

the NGR, would have kept a lot of work from being done in the wrong direction.

Another example involves the relationship between the NGR and the cost per square foot of building as opposed to cost per unit of use. Consider the design of a jail with cells for individual inmates. The efficiency can be improved only up to a point, by decreasing the corridor width to the legal minimum and by decreasing the cell width. Adding to the capacity of the jail by adding more cells with the same dimensions will increase the total cost by the same amount per inmate, and at the same cost per square foot of floor area. However, if instead another bed is added to the cell, the

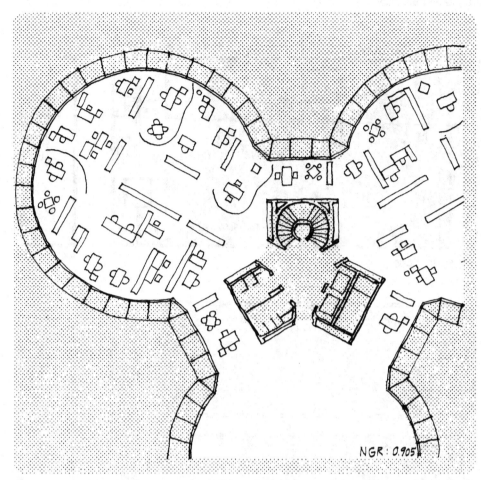

AN EFFICIENT "OPEN LANDSCAPE" OFFICE BUILDING (KIESSLER 77)

NGR: 0.905

FIGURE A1-2. An efficient office floor plan.

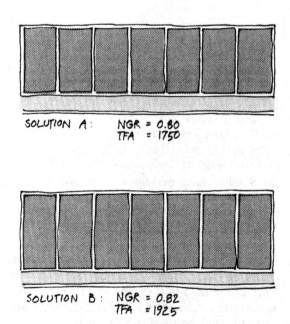

SOLUTION A: NGR = 0.80
TFA = 1750

SOLUTION B: NGR = 0.82
TFA = 1925

FIGURE A1-3. "Improving" NGR through increased total floor area.

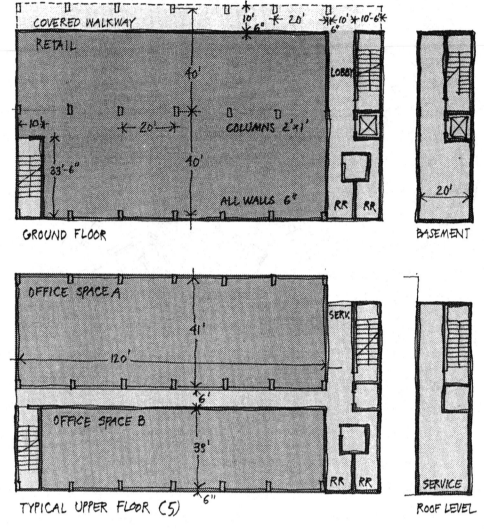

FIGURE A1-4.   Exercise.

increase in capacity can be achieved by only adding a few linear feet to the length of the cell, involving a few linear feet of wall and a few square feet of floor and ceiling area, which makes the cost per inmate significantly lower.

STUDY QUESTIONS

1. For the example shown in Figure A1-4, calculate:
   (a) The net usable area (NUA) per office floor.
   (b) The net rentable area or net leasable area (NLA), assuming several tenants on one floor and no inclusion of common spaces.
   (c) The rentable area if the entire floor is leased by one tenant.
   (d) The rentable area for tenants A and B, respectively, if a prorated share of common areas is to be included.

   (e) The shop area for a retail tenant on the ground floor.
   (f) The total or gross floor area (TFA) for the entire building.
   (g) The NGR values for the assumptions of (b), (c), and (d) above.
   (h) The floor area ratio (FAR).

2. Calculate the NLA as in case (b) above, the TFA, and the NGR for variations of the building:
   (a) Smaller area per floor—five structural bays rather than six—and instead one more floor.
   (b) A bay size of 45′ by 20′ rather than 40′ by 20′.
   (c) A bay size of 35′ by 20′. (All other assumptions remain the same.)

3. What would be the total rent paid by tenant A assuming case (1b) and a rental rate of $20/sf of NLA as opposed to that paid in case (1d)? For the assumptions in question 2? What if the renal rate for case (1d) were reduced to $19.00/sf?

Appendix 2
The Building Delivery Process

INTRODUCTION

The problems of design and decision-making from an economic point of view will look quite different and will require different methods of analysis, depending on whether they occur early or late in the building delivery process. To understand these differences and the proper application of various techniques to these problems, we must understand the process and the nature of decisions made by the various parties at each of its stages. Here we look at the sequence of steps that constitute the process by which buildings are planned, designed, constructed, used, and ultimately disposed of. We discuss the activities that occur at each stage of this process, the participants in the process, and the decisions for which they are responsible. Finally, we explore some options for modifying the traditional delivery process to improve its economic performance.

The building delivery process is narrowly defined as the sequence of activities and decisions followed by various participants including the client or owner and the architect/designer for the purpose of conceiving, planning, and implementing a building construction project. However, to evaluate the economic consequences of these decisions, it is necessary to include in this scheme the phases of use, future modification, expansion, alteration, remodeling, and ultimate disposal through sale or demolition.

THE MAIN PHASES OF THE BUILDING DELIVERY PROCESS

The main phases of the building delivery process are roughly as shown in Figure A2-1. There will always be some repetition in this process; for example, if cost estimates for preliminary design, design solutions, or, eventually, bids turn out over budget, redesign may be needed. Similarly, sequences of use – remodeling – reuse or of use – expansion – use, and the like, could occur repeatedly.

RECURRING QUESTIONS AND ACTIVITIES

Throughout the process, a few major types of activities recur many times in different forms. They are:

- *Planning and design:* This refers to conceptualization, planning, and description of the project — including the more narrowly defined architectural design during the schematic design and design development phases. From inception onward, this activity of planning occurs many times, in increasing detail, for many different aspects of the project, and with different names.
- *Analysis:* This is understood here as the activity of investigating the consequences of proposed design and other planning decisions — for example, rechecking the estimated cost of the building after each round of changing or refining the design (description).
- *Decision-making:* This activity involves selecting from alternative courses of action and making commitments to adopt the results of previous planning and analysis as the basis for subsequent work.
- *Implementation:* This is the actual work of putting the plans into motion: producing the construction drawings, specifications, and other documents; carrying out construction; and so forth.

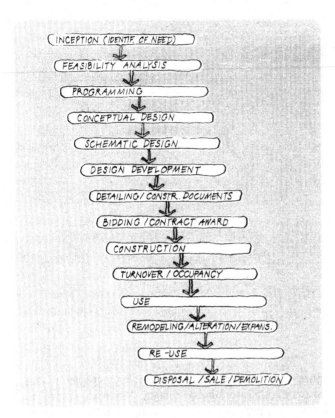

FIGURE A2-1.   The phases in the delivery and life of buildings.

Figure A2-2 shows the major activities and decisions for each of the phases listed above.

PARTICIPANTS

From the list of the above activities and decisions, it should be evident that there are many more parties involved in the process than merely the owner, architect, and contractor. The following is a listing of some of the main groups of participants, including some for which the corresponding activities are not even included in the above list, such as the manufacturers of building materials and components, transportation firms, and so on:

- Client/owner/user (i.e., client who will also use facility): head of household, family members, investor/owner, management, employees.
- Client/owner (i.e., not necessarily using facility): investor, manager.
- User: residents/shop/office tenants/students/patients/visitors/shoppers; staff/employees, including worker, personnel, janitorial, security staff, and so on.
- Architect (including several specialty roles within the office)
- Programmer
- Market analyst

- Feasibility analyst
- Engineers
- Financing/lending institution (bank)
- Developer
- Contractor
- Subcontractors
- Real estate taxing and appraising agency (county, city)
- Utilities
- Manufacturers
- Transportation firms/industry
- Labor
- Government (federal; state; local): regulatory, financing/subsidizing, permitting agencies
- Insurance agent
- Public
- Realtor

For a number of reasons, it is not possible to assign the participants to their respective activities and decisions in a straightforward manner that would be valid for all cases and building tasks. For example, the process for the planning and construction of a single-family home and that for a speculative commercial office building and the financing and procurement of a building by a government agency will be very different, not only in the distribution of responsibilities but also in the sequence of activities. Some of the factors that determine the specific arrangement in a given situation are discussed briefly below. It is a good idea at the start of each specific project to outline the anticipated delivery process, complete with all participants and their respective responsibilities and roles, and to plan actively for constructive interaction and cooperation among them. All these participants, whatever their role, also are economically involved in the process, and will pursue, among others, economic objectives with respect to the planned project. The various objectives are sometimes in conflict. If this is not acknowledged, the objectives are not understood, and no provisions made for their cooperative resolution, disruptive behavior and other problems are likely.

VARIATIONS OF THE PROCESS

Major differences in the building delivery process are due to the nature of the owner and owner objectives regarding the project: Is it for owner's own use or for profit? Is the owner a governmental agency, building for its own use or that of another governmental entity? The source and process of funding and ultimate payment will introduce additional differences. This is particularly acute in the case of governmental projects or projects involving building types that require considerable governmental regulation and/or subsidies. Health care facilities and educational facilities are examples of

PHASE	ACTIVITY	DECISION
INCEPTION	Conception	
	Determination of need for project	Go ahead?
		Commit resources for feasibility studies?
FEASIBILITY	Analysis of market, comparable projects.	
	Estimate of probable costs, income, benefits.	Determination of budget
	Feasibility analysis: can it be built for	
	available budget? Will it be competitive?	Feasible?
	Establish site selection criteria.	Adopt criteria
	Analysis of potential sites	Site selection
	Possible financing	Financing plan
	Develop project delivery plan	Project delivery plan
		Selection of programmer
		Go ahead?
PROGRAM	Analysis of need: activities, users, functions	
	Preparation of program: space requirements	
	Cost estimate; check against budget	Still feasible?
	Feasibility analysis	
	Formulation of program	
	Site analysis	Site recommendation
	Project delivery plan, schedule, contractual	
	arrangements	Adoption of financing plan
	Arrangement of financing	Go ahead?
		Adoption of program
		Adoption of project plan
		Adoption of schedule
		Site acquisition
	Negotiation w. architects / Selection of architect	Setting budget
		Commissioning architect
DESIGN/PREPARATION OF CONSTRUCTION DOCUMENTS		
Concept. design	Prepare/present concept	
	Arrange contracts w. engineers/consultants	
	Site / soil tests	Go ahead?
Schem. design	Prepare/present schem. design	
	Analysis: still feasible? Within budget?	Adopt schematic design
	Coordinate consultants' work	
Design development	Preparation of consultants' reports	
	Analyze consultants' reports	Adopt reports / results?
	Prepare/present design	Still feasible? Within budget?
	Preparation of construction documents	Approve design / Go ahead?
	Preparing constr.drawings	
	Specifications	
	Cost estimate/ analysis: within budget?	
	(Redesign)	
		Approve constr. doc.: Bid?
BIDDING / CONTRACT AWARD	Prepare bids	Finalize bid (contractor)
	Analyze bids: Within budget?	Accept/reject bid
		Go ahead? Award contract?
CON- STRUCTION	Prepare construction management plan / schedule	
	Arrange interim (construction) financing	
	Obtain permits	
	Order materials	
	Prepare site	
	Construction	Accept/ approve work
	Construction inspection	Approve payment
OCCUPANCY	Arrange permanent financing	Close perman. financing
	Arrange final inspection	Certif. of occupancy
	Advertise for tenants	
	Arrange / negotiate leases / sale	Sign lease / purchase
		Plan tenant improvements
		Implem. tenant improvements
	Prepare moving plan	
	Moving	
USE	Use of building	Continue / end use/ lease?
	Consider change of use	Approve use change?
	Operation	
	Management	
	Plan / schedule maintenance, repair work	
	Maintenance	
	Repair & replacement / Identify items for rep/repl.	Implement repair/replacement?
	Cleaning / janitorial service	
ALTERATION / MODIFICATION / EXPANSION		
	Prep. plans / schedule for expansion / alterations	
	Identify need for expansion / alteration	Go ahead?
	Implement alteration / exp. plans	
SALE	Advertising for sale	
	Analysing property	
	Prepare offer / Arrange financing	Make offer?
	Analyze offer	Accept / reject offer
DEMOLITION	Identify necessity for demolition	Go ahead?
	Implement demolition	

FIGURE A2-2. Activities and decisions in the phases of the delivery process.

this. Both have complex, lengthy review and approval processes with severe budget constraints, but the process is quite different in each case.

The differences in the process and guidelines for developing the program will be responsible for further modifications. For example, is the program budget-driven (i.e., by the allotted budget or amount of equity investment available to the owner), or is it determined by needs criteria and facility program guidelines governing a building type regulated (and funded) by a government agency? These differences are of particular interest to the student of building economics because they determine owner/investor objectives and behavior in ways that may appear inexplicable to those who are not aware of the particular conditions. For example, many government buildings are financed by a capital improvements funding process that governs only the initial cost (perhaps through legislative action), with the funds for operation and maintenance of the buildings coming from an entirely different budget. This arrangement often removes the incentive for planners to seriously consider meaningful trade-offs between the initial cost and possible long-term savings for the agency re-

sponsible for the initial cost—even when there also are legislative requirements that a life cycle cost analysis be performed for such projects.

Fast-Track and Design–Build Project Delivery

Finally, various efforts to improve both cost control and speed of delivery through the contractual arrangement between, for example, the architect and the contractor have led to changes in the sequence and allocation of responsibility of many activities. The two best known approaches in this respect are the Fast-Track method and Design–Build.

In the fast-track approach, construction contracts for parts of the project are bid and awarded as soon as the construction documents for the respective parts of the building are completed, so that construction can begin while other parts still are being designed and specified. This results in an overall reduction of the delivery time due to overlapping or telescoping of the planning and construction activities. The drawback is that no final cost figure can be established when construction starts, and many clients do not like the resulting uncertainty

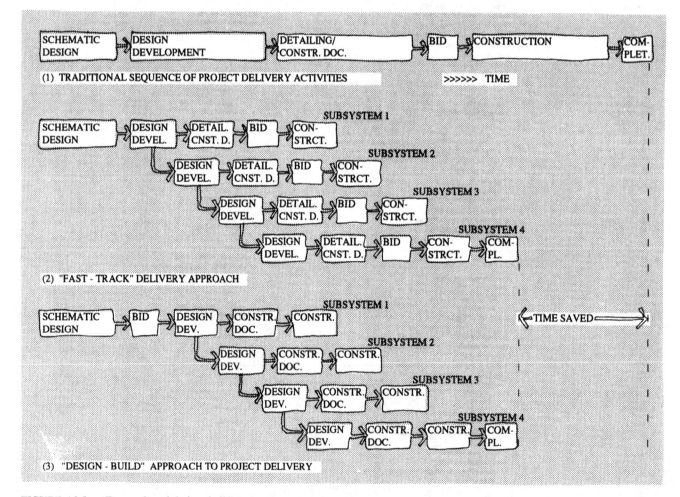

FIGURE A2-3.   Fast track and design–build approaches to building delivery.

TOTAL INITIAL PROJECT COST - DATA		ALTERNAT. A	ALTERNAT. B		
	Site area	SITAR	100000	100000	sf	Given
	Site price	SITPRC	4.5	4.5	$/sf SITAR	Context
	Demolition cost	DEMCST	$5,000	$5,000	$	Estimate
	Parking requirement	PARQ	200	200	sf of NLA/park. sp	
	Parking space requ.	PARSPQ	280	280	sf/car(incl. driveways)	
15)	Parking spaces needed	PARKSP	250	250	#	NLA/PARQ
14)	Needed parkg. &drives	PARKAR	70000	70000	sf	PARKSP*PARSPQ
	Parking&drive price	PARPRC	4.25	4.25	$/sf of PARKAR	
	Building footprint	BFP	9209	9209	sf	FA
17)	Landscaped area	LSAR	20791	20791	sf	SITAR-BFP-PARKAR
	Landsc. price	LSPRC	5.00	5.00	$/sf of LSAR	
	Runoff retention cost	RETCST	1500.00	1500.00	$	Estimate
	A/E fee percentage	AEFEEPC	10.00%	10.00%	% of BLDCST + SWCST	
	Permits percentage	PERMPC	2.00%	2.00%	% of BLDCST + SWCST	
	Legal fees percentage	LEGFEEPC	1.00%	1.00%	% of BLDCST+SWCST+SITCST	
	Variance appeal cost	VARCST	$0	$0	$	Estimate
	Development costs	DEVCST	10000	10000	$	Given
	Accounting fee perc.	ACCFEEPC	1.00%	1.00%	% of BLDCST+SWCST	
	Surveying cost	SVCST	$2,000	$2,000	$	Estimate
	Utilit. hookup length	UTHKL	75	75	lf	
	Util. hookup fee	UTHKFEE	60	60	$/lf util.run	
	LCC Analysis fee	LCCAFEE	$1,500	$1,500	$	Estimate
	Constr. contingency%	CCONTPC	5.00%	5.00%	% of BLDCST+SWCST	
	Soil tests fee	SLTSTCST	$1,500	$1,500	$	Estimate
	Fixed equipment cost	FEQPCST	$0	$0	$	Estimate
	Furniture cost	FURNCST	$5,000	$5,000	$	Estimate
	Carrying charges:					
	Real est. tx. rt. (constr.per.)	MILLS	2.80%	2.80%	% of SITCST	Context
	Site security cost/mo.	SITSECST	$750	$750	$/mo of constr.	Estimate
	Insurance (constr.)	SITINS	$300	$300	$/mo. of constr	Estimate
	Moving cost	MOVCST	$0	$0		Estimate
	Constr. loan LVR	CLVR	0.8	0.8		Context
	Constr. loan interest rate	CLINT	12.00%	12.00%	%of CLOAN/yr	Context
	Constr. loan AVOBPC	AVOBPC	50.00%	50.00%	% of CLOAN	
	Constr. Loan Points	CLDSCPTS	3.00%	3.00%	% of CLOAN	
	Mortgage Loan Points	MLDSCPTS	0.35	0.35	% of MLOAN	
	Mortgage LVR	MLVR	0.8	0.8	% of MLOAN	
	Construction period	CP	18	12	months	

TOTAL INITIAL PROJECT COST - AREA METHOD - CALCULATIONS					
11)	Site cost	SCST	$450,000	$450,000	$	SITAR*SITPRC
13)	Parking & driveway cost	PARCST	$297,500	$297,500	$	PARKAR*PARPRC
16)	Landscaping cost	LSCST	$103,953	$103,953	$	LSAR*LSPRC
12)	Site work cost	SWCST	$402,953	$402,953	$	PARCST+LSCST+RETCST
18)	A/E Fees	AEFEES	$528,393	$528,393	$	AEFEEPC*(BLDCST+SWCST)
19)	Permit costs	PERMCST	$105,679	$105,679	$	PERMPC*(BLDCST+SWCST)
20)	Legal fees	LEGFEES	$57,339	$57,339	$	LEGFEEPC(BLDCST+SWCST+SCST)
21)	Accounting fees	ACCFEES	$57,339	$57,339	$	ACCFEEPC*(BLDCST+SCST+SWCST)
22)	Utility connection cost	UTCST	$4,500	$4,500	$	UTHKL*UTHKFEE
23)	Construction contingency	CCONT	$264,196	$264,196	$	CCONTPC*(BLDCST+SWCST)
25)	Property tax (dur. constr.)	PROPTXC	$18,900	$12,600	$	(MILLS/12)*SCST*CP
26)	Site security cost	SECCST	$13,500	$9,000	$	SITSECST*CP
	Insurance cost(constr.)	INSCST	$5,400	$3,600	$	SITINS*CP
24)	Carrying charges	CARCH	$37,800	$25,200	$	PROPTXC+SECCST+INSCST
10)	Project cost (to be financed)	PRJCST	$6,809,174	$6,796,574	$	BLDCST+SCST+SWCST+AEFEES+
33)	Construction loan	CLOAN	$5,447,339	$5,437,259	$	PRJCST*CLVR
32)	Constr. loan fee	CLOANFEE	$163,420	$163,118	$	CLOAN*CLDSCPTS
34)	Constr. loan interest amnt	CLINTAMNT	$490,261	$326,236	$	CLOAN*AVOBPC*(CLINT/12)*CP
31)	Constr. Financing cost	FINCST	$653,681	$489,353	$	CLOANFEE+CLINTAMNT
30)	Total Initial Project cost	TIPRJCST	$7,462,855	$7,285,928		PRJCST+FINCST
	Total cost per sf TFA - Area method		$81.04	$79.11	$/sf of TFA	TIPRJCST/TFA
	Total cost per sf NLA - Area method		$149.26	$145.72	$/sf of NLA	TIPRJCST/NLA

FIGURE A2-4.   Effect of delivery time on initial cost.

with regard to the project budget. The design–build approach remedies this by using a combined architect–contractor team (design–build team) that gives the client a final contract at the outset. The team then proceeds with the planning and construction work as in the fast-track approach.

Both approaches require that the contractor be brought in early in the planning process, both work best with standardized construction methods or building systems, and both leave the client with some uncertainty as to the final product at the time when the crucial decision to proceed with the construction must be made. They require that the owner have a higher degree of trust and confidence in the architect and the contractor than would be necessary if all the work were carefully specified and documented in working drawings prior to signing of the contract. In times of high interest and high inflation rates, the initial cost savings due to the reduction in delivery time may be well worth the potential drawbacks.

SUMMARY OBSERVATIONS

The building delivery process, as described by its sequence of steps, activities, and the decisions made by the various participants, can influence not only the economics of the building but also the economic objectives of the parties involved and thus the analysis techniques that should be used to investigate the appropriateness of design solutions. In each particular situation, then, the probable process should be carefully studied and understood.

A few main observations can be made:

- Each phase of the process uses the results of previous phases as starting assumptions.
- With each change in assumptions and decisions about program and design, the implications on the project's economic viability should be checked again; feasibility analysis should accompany the entire process at each step.
- Each time that the process must revert to a previous step—either because its assumptions have proved to be inadequate, or because assumptions and objectives have changed—the repetition will cost both

time and money. It is also likely to cost the responsible party credibility and reputation.
- Care should be taken, therefore, to arrange the process in as straightforward a manner as possible, ensuring that the analysis at each step will result in reliable assumptions to serve as the basis for the next steps, avoiding the need to go back to a previous stage.

In other words, the delivery process must be considered as part of the design problem; it must itself be carefully planned.

STUDY QUESTIONS

1. In the spreadsheet for initial project cost (shown in Figure A2-4), shorten and lengthen the construction period (CP) by 10%, 20%, and 50%, respectively, and check the effect of these changes on the initial project cost if only the financing charges on the construction loan (10% interest, 2% discount points) and carrying charges ($500/month) are considered.
2. Make the construction price a function of time due to inflation (say, INFL = 6% annually) by inserting a figure for inflation, and calculate the actual construction price as $CPR_a = CCPR_o*(1 + INFL/12)^{CP}$. Analyze the resulting effect of variations in the construction period on initial project cost.
3. Cost estimates during the feasibility and programming stages have used average assumptions for the construction price (say, $60/sf of TFA at the date of the estimate). To this 6% inflation for the construction period (1 year) and the remaining planning period (6 months) would have to be added, as well as construction financing costs. Based on experience, you estimate that the bidding process in the traditional delivery method will result in low bids that are about 5% below this figure. It is proposed that a design–build process be adopted instead, which will use the average cost plus inflation as the basis for the contract but promises to cut the entire delivery time until occupancy down to 9 months. Which delivery method would be more advantageous to the owner? By how much would the delivery period have to be adjusted for the alternatives to be equivalent?

Notes on Measurement

MEASUREMENT PROBLEMS; KEY CONCEPTS

Much of building economics can be seen as an effort to establish sufficiently precise and useful measurement about costs and benefits of building projects — precise enough to give decision-makers the confidence to distinguish between good and bad solutions. Precise measurement has contributed significantly to the success of science (or, more specifically, the "hard" sciences) during the past centuries — some say science *is* measurement. Because of this importance, arguments in everyday discourse that are supported by numbers resulting from measurement tend to carry more weight with audiences than those that have no such backing — even if the numbers only have the look but not the substance of reliable "scientific" measurement. Unfortunately, this is true more often than we would wish, even and especially in economics, which likes to pride itself on relying only on "cold" facts and "hard" numbers. Where value — that is, judgments about value — comes into play, many of the assumptions we make about the nature and reliability of measurement in, for example, the physical sciences no longer apply. This appendix serves as a brief review of some of the fundamentals of measurement.

Definition of Measurement; Unspoken Rules and Conventions

The dictionary defines "measurement" as "determination of the magnitude of a quantity by comparison with a standard for that quantity. . . . To express a measurement, there must be a basic unit of the quantity involved" (*New Columbia Encyclopedia*). This requires some elaboration.

Implied in the stated definition is the idea that there is an entity that can be described with an attribute — a variable — that can take on different quantitative values. Briefly, what is measured will always be a variable, and it should be a habit to specify and name that variable explicitly. For example, the length of a building is a variable (i.e., buildings vary with respect to length); the length of the building in which I work is 65 feet. Thus "65" is the value of the variable "length" for this building. "Foot" is the unit of measurement that is counted or estimated. The encyclopedia makes a distinction between counting (the result of which is always exact because it involves discrete entities not subdivided into fractions) and measurement; the latter is always an estimate because it involves smaller and smaller fractions of the unit. As we will see, this view of measurement is somewhat narrow; a wider sense might simply focus on the establishment of the value of a variable. This would allow us to include such notions as the determination of the color, or the material of something, in the meaning of measurement. The difference lies in the nature of the unit of measurement involved. "Blue" is a value of the variable "color" — but what is the unit? In this case, it is simply the name of a color.

Scales of Measurement

To account for these differences, there are four types of scales of measurement: the nominal scale, the ordinal scale, the interval or difference scale, and the ratio scale.

The nominal scale involves simply distinctions of values of a variable that are merely named but are not expressed in a quantitative manner or arranged in any particular order. Examples are the variable "color," already mentioned (with the values being the various colors: "blue," "red," etc.), or "material of wall," or "window type." In some cases, numbers themselves are mere values of nominal scale variables, such as telephone numbers or social security numbers. They only identify or name individuals (the term "nominal" derives from the Latin *nomen* for "name"). The numerical value of these numbers is of no significance.

The ordinal scale also makes distinctions among individual entities, but in addition arranges them in a definite order. The class terms for student— "freshman," "sophomore," "junior," "senior"—are examples of values of a variable on an ordinal scale. The Mohs scale of hardness of materials is one of the best-known examples. It assigns numerical values to materials (the diamond as the hardest substance is assigned 10, talc as the softest is given 1), but the numbers are based only on whether one material scratches another. This allows the materials to be arranged in a definite order. The numbers express that order but otherwise do not mean anything; it would be inappropriate to infer, for example, that diamonds (10) are twice as hard as apatite, which has a hardness number of 5. For this reason, one cannot perform mathematical operations on "measurement" results of variables on an ordinal scale. Perhaps we should say "one should not" because this methodological mistake is quite widespread. Ironically, the academic world, which uses letter grades to express the performance of students, is the worst culprit with its practice of not only assigning numbers for the letters ("4" for "A," "3" for "B," "2" for "C," "1" for "D," and "0" for "F") but then to form grade-point averages from these numbers—as if "A" were indeed twice as good as "C."

Both the nominal and the ordinal scale do not involve standardized units of measurement other than the name of the value, which is why many people do not acknowledge measurements on these scales as true measurement. This changes with the difference or interval scale and with the ratio scale. Here, the measurements are expressed by means of giving the count (at least) or count plus fractions of the standard unit. The choice of unit is a matter of convention, that is, it is essentially arbitrary. Length can be measured in feet or meters or cubits, temperature in degrees Celsius or Fahrenheit.

What does matter significantly is whether the units are such that they have an arbitrary or natural zero point. This difference distinguishes measurements on the difference (interval) scale from those on the ratio scale. Only the latter has natural zero points; zero length is zero in all systems of measurement. This is not so with the difference scale, as the temperature scales mentioned demonstrate. There, the zero point is also arbitrary and only defined and fixed by convention.

Zero degrees Celsius is defined by the melting point of ice, whereas Fahrenheit (so the story goes) simply used the coldest temperature he recorded in his home town as the zero point. The usefulness of these reference points as measurement conventions rests in the extent to which they represent constants in nature. The freezing and boiling points of water are such constants (given certain conditions of water purity, atmospheric pressure, and so on). Similarly, the human body temperature, which served as Fahrenheit's other reference point (100 degrees) is almost a constant; ironically, Fahrenheit must have run a fever on the day he decided to calibrate his scale.

The practical implications of these differences are that only with measurements on the ratio scale are all mathematical operations (addition, subtraction, multiplication, division) permissible and meaningful. For measurements on the difference scale, most operations are meaningful only when performed on ranges or intervals of values (hence the name "interval scale"). For example, $[212^*F - 32°F] = [100°C - 0°C]$; but the midpoints between these ranges: $[212°F - 32°F]/2 = [100°C - 0°C]/2$ are $90°F$ and $50°C$, respectively, which does *not* mean that $90°F = 50°C$; these values must first be added to the starting points of the ranges: $90°F + 32°F = 122°F$, which is equivalent to $50°C + 0°C = 50°C$.

Some of the measurements we are concerned with in design and in building economics have the appearance of ratio scale variables and often are treated as if they were, but in reality are only ordinal or difference scale measurements. The problems arising from this can be significant.

Objective versus Subjective Scales; Objectification

Additional problems arise from the common expectation that measurements should be objective rather than subjective. This means that measurement results should, objectively, reflect the true property of the thing or object measured, regardless of how the person reporting the measurement may, subjectively, feel about it. Scientists must insist on this standard, and have instituted elaborate procedural safeguards to ensure the objectivity of their findings. Measurements must be taken according to a systematic, explicitly described procedure; the procedure must permit anyone, anywhere to replicate the conditions under which experiments and measurements were taken, to check whether he or she will obtain the same results. To achieve this, the entire process must be made available to anyone through publication.

Many of these provisions are the subject of implicit assumptions. For example, the measurement of the width of a room is tacitly assumed to adhere to the rules that it is taken perpendicularly to the walls, that care is taken to keep the measuring tape straight and taut, not to let it sag, that it be defined whether the width should mean the width between the walls or the width between the baseboards, and so on.

The ideal of objectivity in science has been the subject of some controversy, starting with the observation that many measurements are estimates—the result of procedures using standards that in themselves are not objective but mere conventions (e.g., taking the average of a set of measurements as the accepted result). The controversy is especially bitter in connection with attempts to apply standards and methods from the natural sciences to the social sciences.

Very clearly, problems arise when the entities measured are in themselves matters of value placed on things (buildings, etc.) by persons, that is, matters of subjective judgment, by definition. For one thing, the scales upon which judgments are expressed usually are not ratio or even difference scales, but typically are ordinal scales. Their calibration and interpretation (i.e., the definition of units of measurement so that it means the same thing for everyone involved) pose great difficulties. Rather than being acknowledged, confronted, and resolved, however, the problems often are sidestepped. This process often takes the form of using precise-looking measurement scales for ordinal scale judgments and subjecting these judgments to elaborate statistical procedures giving precise-looking results that obscure the fact that the measurements (judgments) do not have the same meaning for any two people involved.

One way to acknowledge these problems would be to accept the fact that judgments about values are and will remain subjective, but to concentrate our efforts on the mutual explanation of the basis of such judgments. Sometimes we can succeed in showing each other how a subjective judgment depends on some objective property of the thing evaluated. This process is called "objectification"—not the most attractive term, and not to be confused with the attitude of "turning people into objects" (which some opponents of the systematic treatment of problems involving subjective judgments sometimes allege to be the sinister intentions of the scientific community).

GUIDELINES FOR DEALING WITH MEASUREMENT

The reasons for discussing these concerns in a book on building economics can be summarized in a few guidelines for dealing with measurement so as to avoid some of the more pernicious pitfalls. Briefly, we should keep the following points in mind and, as necessary, be sufficiently specific in providing definitions and explanations:

- Name the variable explicitly (stating in full the name of the variable measured, adding any necessary explanations).
- For convenience and conciseness especially in writing equations, use symbols that stand for the variables. It is a good idea to use mnemonic symbols—abbreviations of the variable name that still remind us what the symbol stands for.
- Specify the unit of measurement by referring to the standard used, or by defining such a standard.
- Be aware of the scale of measurement on which the variable in question is measured, and use mathematical operations appropriate to that scale. Specifically, be aware of the nature of the unit of measurement (judgment or physical property? natural or arbitrary zero? numerical values that refer to defined quantitative standards or to comparative judgments?).
- Specify the procedures involved in the measurement. Are they in turn objective, and can they be described systematically enough to be reproduced by others to give the same results under the same conditions?
- Be aware of the various conditions under which the measurements are or can be taken, their potential influence on the readings, and the possibility of in turn describing these conditions objectively and systematically.

MEASUREMENT ISSUES IN BUILDING ECONOMICS; PROBLEMS WITH THE USE OF MONEY AS A YARDSTICK OF VALUE

With respect to measurement, building economics poses some challenging conceptual problems. Much of its main effort is directed at the task of developing (theoretically) appropriate measures of performance for building projects, and then ascertaining the values of those measures for proposed project alternatives in practical application. "Ascertaining" usually is understood as "measuring," but in building economics, it more commonly means "estimating." Estimates are fraught with the possibility of error. The kinds of measures we are concerned with all are related to the future; and even where we have solid measures from the past and present to work with, the task of deriving estimates for the future from that information requires the application of a lot of additional judgment—not to say "guesswork"—which compounds the uncertainties in the resulting estimate.

There are even more fundamental difficulties that affect the very concepts of measurement or estimation, having to do with the way we talk about the performance or costs and benefits related to building projects, with the way we lump different costs and benefits together into overall cost and benefit accounts, and with the different kinds of currencies we use to compare the two. More specifically, these questions deal with the appropriateness of using money or any other yardsticks—such as energy, labor, or judgment scores—as a measure of costs and values, and the nature of those yardsticks.

Money as a Yardstick of Value

The very concept of economic performance, together with the practice of using performance measures (variables) that either themselves are money amounts or are constructed from several such money amounts, suggests that money constitutes a valid, objective tool of measurement of cost or value, which is given a dollar price tag. Many people do not see the problem here, or the source of confusion, at least with respect to costs. In this view, the cost of a building is the amount of money we must pay to get it built; the money amount is an objective measure, so cost is an objective measure. Things are not quite so simple with regard to measures of value (benefits; at least once we go beyond the straightforward monetary benefits of rental income or sale price), which are more easily seen as subjective judgments.

If, however, the money amounts for cost still are held to be objective amounts, serious problems arise with regard to any measures of performance that are composed of several money-related variables (e.g., the Benefit–Cost Ratio). Are they objective or subjective? A crucial distinction must be made to resolve these problems: cost and benefits and money amount are different variables.

If it is seen as a physical entity (i.e., a stack of dollar bills), the idea that a money amount is an objective variable, accurately measured by a count, is certainly valid. It meets all the criteria for objective measurement: it is a variable using a ratio scale with a natural zero (while the unit can be arbitrary); a procedure is used to establish the measurement such that anybody can use it to get the same results no matter what his or her feelings; it permits all kinds of mathematical operations; and so on. However, both cost and benefit, or value, are not only composite entities made up of many separate considerations, but they also are entirely subjective. Once we make this distinction, we can raise the question of how we get from such subjective judgments to the supposedly objective dollar amounts, and what common yardstick should be used for such evaluations.

Several such common yardsticks have been suggested; of these, money is by far the most prevalent. It is used to form more complex measures of performance, such as the Benefit–Cost Ratio. The possibility of mistaking the objective measure of money (as a stack of dollar bills) for the subjective measures of money as the yardstick of cost and value is then transferred to these measures as well. An alternative is to use some judgment scale to express value and cost judgments instead.

For comparisons of costs of living and living commodities, hours of work have been used, such as the number of hours a worker must work for a house, or a car, in different countries. Finally, energy has been suggested as a common currency alternative to money.

Even the valuation of money itself is subject to additional distortions and deviations from a straightforward ratio-scale measurement. This further erodes its status as an indisputable, apparently objective measure of value.

FACTORS INFLUENCING THE VALUATION OF MONEY

Some of the main sources of distortion of the objective reliability of money as a measure of value will be discussed in the following paragraphs:

- The decreasing marginal utility of both money and the units of goods it measures.
- The exchange function of money.
- The problem of value judgments that apply not only to goods by themselves but to combinations of different amounts of goods (the problem of "adding apples and oranges," as well as the problem of different levels of satisfaction with different combination amounts, between which consumers would have to choose if limited amounts were available).
- The problem of intangibles.
- Distortions of market prices due to the laws of supply and demand, differences in costs of production, competition, taxation, and subsidies.
- The difficulties resulting from both charging costs and the distribution of benefits for private, collective, and quasi-collective goods, respectively.
- The effect of uncertainty and risk in the assessment of costs and value benefits we expect to occur in the future.
- The influence of time in the assessment of value, including the value of money. (This subject is so important that it is treated separately, in Appendix 8.)

Decreasing Marginal Utility

Formation of the benefit–cost ratio, as one example of a measure of performance formed by combining several money measures, rests on the unspoken assumption that it is legitimate operation—that is, that money, or more specifically, the value expressed by a money figure, is measured on a ratio scale.

If value measured in terms of money were indeed on a ratio scale, the following would have to be true: the value placed by somebody on one additional unit of money obtained should be the same whether that person already had many units (dollars) or none. In reality,

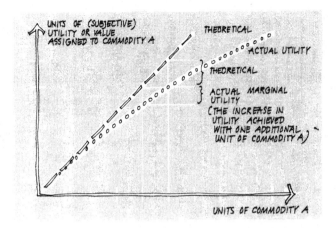

FIGURE A3-1.   The decreasing marginal utility of money.

this is not the case—with money or with many other things for which a straightforward, one-for-one equivalence of value for money units is assumed.

This phenomenon (that we place less value on one additional unit if we already have a lot than if we have nothing) is that of decreasing marginal utility.

The Exchange Function of Money

Money is used as the medium of exchange for goods when anything is bought or sold. The assumption of a "free market" includes the notion that we can bargain about the price of something until we feel that the price is fair: that we get adequate value for the money, and conversely, that we obtain an adequate amount of money for what we sell.

Closer reflection reveals that we cannot assume that the price paid is a precise measure of the value received, even if the transaction was the result of a bargaining process. A deal is struck because the seller of goods wants the money more than what is sold. Conversely, the buyer wants the item bought more than the money paid. If this were not so, nobody would bother to "come to the market." The difference in value placed by each party on what is received and what is given in return must at least exceed the effort involved in setting up and bargaining for the deal. Persistence in bargaining may bring the price closer to the actual value placed on something, but some slack must remain. Indeed, the better the deal as perceived by one or both parties involved, the greater the value–price discrepancy. This is true even if we assume that both parties generally have the same value system, in that they have the same preferences and priorities. Just saying this is tantamount to reminding us, of course, that people have widely different values. But this is contrary to what the pervasive use of (average? predominant?) money prices for goods appears to suggest. The statement that "The price of a barrel of oil is x dollars" does not mean that

its value to a person A is equivalent to the value placed by A on the amount of x dollars, or that its value to person B is the same as to A.

Combination of Goods

At the very simple level of comparing how people value combinations of two types of goods, economists visualize different value systems by means of the technique of constructing indifference curves. These curves reflect levels of equal satisfaction (i.e., subjective value) for different combinations of good, where there is a limit on how much of each a person could have. For example, Figure A3-2 shows how two people, A and B, might feel about extra patio space and extra storage space in their house, when the budget does not allow for more than a fixed amount of extra space to be provided. It could be given entirely to patio (x), or to storage (y), or divided between the two.

Levels of equal satisfaction or utility are plotted such that a person would be equally happy with any combination of space that lies on the indifference curve. Economists postulate that the "best" combination of both types of space is that where the line representing the cost limit of affordable combinations ($x–y$) touches the highest indifference curve. For A (in Figure A3-2) this would be at A*; for B it would be at B*.

Intangibles

A common objection to the use of money as a performance measure is that we cannot, or should not, "put a dollar tag" on intangible values. Tangible items are

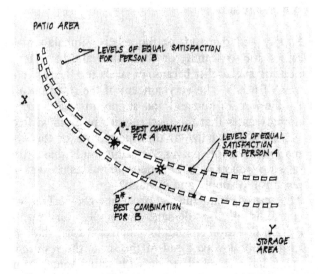

FIGURE A3-2.   Indifference curves. The same budget increment buys y square feet of patio area or x square feet of additional storage area, or any combination indicated by the line $x–y$. The "best" combination for any given person is the point where the line $x–y$ touches the highest indifference curve for that person.

those we can touch (put our hands on), trade, own, and so on. Examples are money, vegetables, clothing, buildings, and so forth. Benefits or values that we cannot touch, trade, own, and so on, but only feel are called intangibles. Examples of intangibles are the stylishness of clothing items, happiness, health, beauty, moral values, life—or the image of a building.

Because of uneasiness in expressing the value of a person's life in terms of money, for example, many persons oppose the use of money scales to express values concerning any intangibles. Proponents of benefit–cost analysis would answer that, whether we like it or not, admit it or not, we are placing dollar tags on such values all the time, even on human life. If the U.S. Congress fails to pass a bill for a certain amount of funds—say, x dollars—that would save y coal miners' lives, it is in effect saying that it (i.e., Congress, but by extension we or the society) places a value of less than x/y dollars on the life of a coal miner. All that Benefit–Cost Analysis does, these proponents would say, is to confront us with these implications of our decisions. We should acknowledge them to consciously deliberate about, discuss, and negotiate them, and not make decisions by default, having closed our eyes to these issues.

Distortions of Market Prices

Everyday experience shows that prices quoted in different corners of life's marketplace do not always reflect the value we place on them, or, better, the value we would have placed on them as the result of an ideal, free market bargaining process. Instead, we are quite used to the notion that prices are less a function of value placed on goods than of a number of market mechanisms that distort price structures. Some of these mechanisms are:

• Supply and demand, or the relative imbalance of prices due to scarcity or abundance and to restrictions on availability because of needs and the seriousness of needs. The very concept of the (theoretical) equilibrium of demand and supply in general economics states that the market price in an ideal free market will stabilize at the intersection of the demand and supply curves. Does this imply that price does not reflect value? Or that our values change in response to supply and demand?
• The cost of producing and procuring goods. This, of course, is related to demand and supply. But consideration of extreme conditions makes it clear that, at least at the low-cost end of the scale, the resulting prices have very little to do with actual value, even when the good in question is a life necessity. For example, clean water is incredibly cheap in most places, considering that life would be impossible without it. Similarly, air is (still) free, both because it is abundant even if not always clean, and because it is

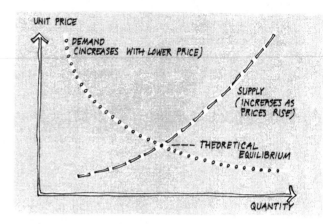

FIGURE A3-3.   Demand and supply.

difficult to control supply. (But already, oxygen is being sold on the streets of some very polluted cities e.g. in Japan).
• The effect of the presence, intensity, or absence or competition. One extreme is the total absence of competition, or a monopoly, in which the sole provider of a commodity is free to charge any price. The other is "cutthroat" competition, which can lead competitors to sell goods at prices well below their own cost just to stay in business, or to force others out.
• The effect of taxation and subsidies. Taxes make some goods artificially expensive for the consumer; subsidies make goods (seem) deceptively cheap because part of their true cost is being paid by the public (i.e., taxpayers in general). Extreme forms of this kind of distortion occur in totally planned economies where all prices are set by government.

Private, Collective, and Quasi-collective Goods

Another question confusing the assessment of value in money terms is the ownership relationship of the things paid for. The assumption of equivalence of money and what it pays for rests on the picture of an exchange of money for goods that then are owned by the buyer. The buyer can use or control the item as he or she pleases, and this right excludes others from doing the same with it. But not all transactions involve goods that have these features. We distinguish:

• Private goods that have the described features.
• Collective goods.
• Quasi-collective goods.

Collective goods usually are bought and paid for by agencies representing all or part of society in general. They are then—unlike private goods—used by all or most of society's members. Examples are roads, public utilities, most government services, and so forth. It is difficult, if not impossible, to deny their use even to those who did not pay their share for them, or even may have opposed their acquisition. With respect to the issue of money and value, this means that goods or values are being paid for that are not appreciated in the same way by all those in whose name and with whose contributions they are bought. Some of these contributors may even consider them disvalues.

To further confound things, some goods have all the features and are being treated, to all appearances, as if they were indeed private goods, but are subsidized by society as if they were collective. Health care and education are examples; individuals pay (at least in part) for, for example, a university degree, earn it to the exclusion of others, and reap its benefits as individuals in the form of increased earnings, prestige, and so on. The underlying rationale is that the public as a whole is better off by having trained engineers, doctors, architects, and so on, around, and thus should support their training. Such goods are called quasi-collective goods.

The issue here is not whether public subsidies to private goods are justified or desirable. What we are interested in is the relationship between value received and money paid for it. With collective and quasi-collective goods, this relationship becomes muddled at best, and violently controversial at worst.

Risk and Uncertainty

Most of the time, our evaluations have to do with plans for events that are expected to happen sometime in the future. But nothing in the future is absolutely certain; there is always an element of risk or uncertainty involved in our plans. We speak of risk if we can make some estimate about the probability (likelihood) of some outcome, and of uncertainty if we cannot make any meaningful probability estimate—perhaps because the outcome of a given action depends on the choices and actions of some other player, who in turn tries to anticipate our own moves.

We find that the attractiveness of the outcomes of alternative actions or solutions changes with our perception of the amounts of risk and uncertainty involved, even with our own willingness or reluctance to accept risk and take chances. Decision theory and game theory are branches of mathematics that deal with the analysis of situations involving risk and uncertainty, and try to develop strategies and decision rules for action in such situations. This is discussed in more detail in Appendix 11.

Time

As indicated above, one of the most important factors that influence how money is valued is the effect of time. The key concepts here are interest, inflation, depreciation, and appreciation. Appendix 8 is devoted to the discussion of these issues and their technical treatment.

STUDY QUESTIONS

1. For each of the following items or statements, indicate whether it is or involves:
   - A variable (if so, also identify the scale and unit of measurement).
   - A value of a variable (if so, also identify the variable, unit, and scale of measurement).
   - A unit of measurement (identify the variable and scale).
   - A subjective judgment or objective measurement.
   (a) The budget for the project set aside by the client is $1,000,000.
   (b) The project is a four-story building.
   (c) It has a net-to-gross ratio of 0.81.
   (d) The project cost should be reduced.
   (e) It will be an award-winning project.
   (f) The last project won third place in the annual state AIA awards.
   (g) The circulation scheme is a double-loaded corridor.
   (h) The cost estimate is $900,000.
   (i) The projected rental return will be above average for this kind of building.
   (j) We should be concerned about the price per square foot for the building.
   (k) The client is worried about the vacancy rate.

2. Why is it important in the scientific method to publish the procedures by which measurements have been obtained? What would it mean for designers to adopt this attitude as a general rule?

3. Are estimates of projected building cost subjective or objective measurements? Why?

4. The "number of beers Joe drank at the bar" is an objectively measured variable. The "value Joe places on a beer" is:
   - An objective measure because he has to pay the same amount of money for each one?
   - An objective measure because each beer bottle contains the same amount of beer?
   - A subjective measure because the marginal utility of the fourth beer is less than the marginal utility of the first?

Discuss your answers. Could the marginal utility become negative?

5. Is the value of a finely proportioned facade:
   - Objectively zero because it does not cost a passerby anything to look at it?
   - Subjectively zero to the passerby if he or she does not notice it?
   - Objectively the difference in cost needed to fine-tune the proportions (as opposed to leaving the facade dimensioned merely the way that the functions called for)? Discuss your answer.

6. Is architecture a private, collective, or quasi-collective commodity?

Appendix 4
Electronic Spreadsheets

ELECTRONIC SPREADSHEET MODELS FOR ECONOMIC PERFORMANCE ANALYSIS OF SCHEMATIC DESIGN SOLUTIONS—OVERVIEW

Conceptually, calculating most economic performance measures for building projects is not very demanding; but it often involves numerous variables and lengthy, repetitive computations. Done by hand or using simple calculators, such analysis in the past tended to be time-consuming and expensive. As a result, standard practice usually required, for example, estimating the initial cost of only one preferred design solution at each stage of design/decision-making. If the expected cost was within the budget, the scheme could be taken to the next stage for further development. Otherwise, the solution would be revised until it achieved an acceptable cost estimate.

Comparison of performance estimates for a larger number of alternative solutions, especially for life cycle cost and other long-term performance measures, was made feasible in ordinary practice only after electronic programmable calculating devices became affordable for the standard architectural office. Such tools—programmable hand-held calculators and computers—not only have reduced the time and the cost of the tasks of estimating costs and performance of proposed design solutions; it is much more significant that they now make it possible to tie the process of estimating economic performance much more closely to the design process itself—that is, they can fundamentally change the character of the design activity.

The "electronic spreadsheet" programs available for personal computers are especially convenient tools for this kind of analysis. Visicalc (the earliest program, now obsolete), Multiplan, Lotus 1-2-3, and Excel are only a few examples of this kind of software, which is widely available today for all brands of computers; and new or upgraded spreadsheet programs are being developed at a rapid pace. There soon will be programs that automatically connect spreadsheets with drawing/drafting software (such as AUTOCAD) and with both specification and cost data bases, so that cost estimates can be taken directly from the drawing as it is being developed on the CAD screen.

The most dramatic effect of combining performance analysis with the design process itself is achieved at the schematic design stage when simple geometric models of the designed object are used as the basis for the estimates.

Geometric design variables, that is, the dimensions of the building, are used as inputs together with context data (unit prices, etc.) adjusted for the geometric enclosure elements of the building. The programs instantaneously recalculate the set of equations connecting the input variables with the performance measures for each new combination of inputs. "Trying out another design solution" now simply consists of changing a few numbers in the spreadsheet and letting the program recalculate the new solution's performance within a matter of seconds. This especially makes the successive-approximation approach to optimization—or better, suboptimization because we are looking at economic performance only—a much more feasible part of the design process. (See Appendix 7 for a detailed discussion of optimization.) Approximation can be done manually, with the designer setting the values of variables for each solution he or she wants to analyze, or automatically by using the iteration feature of most of these programs.

This appendix looks at how spreadsheets can be used for the combined design and analysis of economic performance, which increasingly will characterize the way design solutions are developed and refined in the future.

SPREADSHEET CONCEPTS

Topics discussed in this appendix include: fundamentals of the concept and the use of electronic spreadsheets, and the "top-down" set of equations for the geometric model of a schematic design solution, as opposed to the "bottom-up" representation of those equations in a spreadsheet.

The discussion assumes some familiarity with the following concepts: systems, models, variables, and variable values; design, context, and performance variables; and optimization and the various approaches to optimization, especially successive approximation. Readers should be familiar with the following: the conceptual basis of cost estimating by the area method or "square foot estimating"; the net leasable area (or net assignable area); the gross floor area (or total floor area); the Net-to-Gross Ratio or the efficiency ratio. It would be desirable to already have some experience with spreadsheet programs such as Lotus 1-2-3, Multiplan, or Excel. However, familiarity with the use of this kind of software can be acquired by experimenting with the models shown here and their adaptation to different design solutions. Within the spreadsheet programs, an understanding of the following concepts will be needed, besides the basics of how to start a program, moving around in the sheet, making entries, saving the file, and so on:

- The nature of a spreadsheet as a huge matrix of cells, each of which can be filled with either of the following:

  (a) A content entry.
  (b) Instructions for calculating the value of an entry in a cell from the contents of other cells.

- The different types of entries that can be put into a spreadsheet cells: text; value, formula (instruction), and name or label of the variable in a cell.
- The different ways of referring to a cell in the program:

  (a) By row and column designation (absolute reference).
  (b) By relative reference, by counting rows and columns to the left or right, respectively, above and below a given cell.
  (c) By naming (labeling) a cell and referring to it by that name.

The reader may wish to consult the manual for the software used and to work through the exercises suggested there to become familiar with its application, before starting with the examples here.

At the end of this appendix, upon working through the examples, one should be able to translate a simple problem of calculating a performance variable, first, into a "top-down" mathematical model (i.e, a system of equations); and then into a "bottom-up" spreadsheet model. One then should be able to use the spreadsheet to calculate the performance of a variety of design solutions, which are represented by different combinations of design variables. Trying out different solutions, one could gradually improve the performance of the solutions by successive approximation.

The use of these tools is best explained by means of examples, which one should duplicate in an actual spreadsheet and then use for experimentation. Figure A4-1A shows a diagram of a very simple schematic design solution type for an office building.

We wish to calculate the expected cost for a number of alternative solutions for this building. The steps for setting up a symbolic model for this problem and then transforming it into a spreadsheet will be explained in detail. The reader should follow the steps by working on the computer, and then use the sheet to find a "better" (in this case, better means cheaper) solution. Following this problem some other examples will be shown to demonstrate different aspects of the use of the spreadsheets.

SAMPLE PROBLEMS

The problem calls for the design of an office building with 50,000 sf of net leasable office space, in modules of 500 sf each. We will explore the simple scheme of a building with a double-loaded corridor, lobby and stair with elevator in the center, and fire stairs and other service space at each end (for fire escapes). The modules could have different dimensions (module width and length), and the number of modules on each floor (and thus the length of the building) and the height of

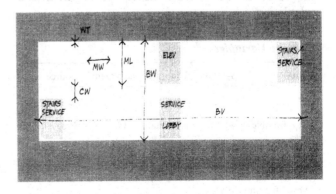

FIGURE A4-1A.   Office building schematic.

the building are open for decision. Figure A4-1A shows a diagram of this simple scheme.

We wish to develop a spreadsheet program that calculates the expected cost of the building, given different combinations of module dimensions and number of modules on a floor, and the resulting building height. The model will represent our current simplified understanding of this problem, which is that the cost depends on the total area of the building (total floor area) and the unit price of dollars per square foot of total floor area. The floor area in turn depends on the net leasable area (called for in the building program) and the additional area needed for corridors, service areas and stairs, elevators, and so on. Of course, this is only one possible way of looking at the problem and of estimating the cost. The model simply reflects this basic working assumption.

Identifying the Variables

As a first step, we should identify the variables we can change (design variables), the variables we cannot change but whose values we have to find out or assume (the context variables), and the performance variable we want to calculate. To establish a habit, which in this simple example may seem excessively pedantic but which later on will become very important, we will list the variables by name, give each a short mnemonic symbol, and identify its unit of measurement [in square brackets]. Sticking to this habit will help us avoid potential confusion and error later on.

Design Variables

Name	Symbol	Unit of Measurement
Module length	ML	[lf] linear feet
Number of floors	NOF	#
Number of modules per floor	NMPF	[#]
Corridor width	CW	[lf]

For the context variables as well as the performance variables, we also need to identify where/how we get the value of the variable: its source or method of calculation.

Context Variables

Name	Symbol	Unit of Measurement	Source
			(generated how?)
Net leasable area	NLA	[sf] square feet	Program
Wall thickness	WT	[lf] linear feet	Standard assumption
(exterior wall only; the interior partitions between modules are ignored for simplicity)			
Module area	MAR	[sf] square feet	Program

Name	Symbol	Unit of Measurement	Source
Construction price	CPRC	[$/sf of TFA]	Assumed; from contractor standard assumption
Floor-to-floor height	FH	[lf] linear feet	

Performance Variable

Name	Symbol	Unit of Measurement
Construction cost	CCST	[$]

Additional (Auxiliary) Variables

We will also need a number of additional variables, the intermediate or auxiliary variables. Although dependent on the design variables, they are not the performance variables we are looking for. They will simplify the process of linking the performance variables with the initial design and context variables, or will serve as performance indicators (see Appendix 9 on performance indicators).

Name	Symbol	Unit of Measurement
Module width	MW	[lf] linear feet
(The module width will be calculated as a function of the module area given in the program and the module length selected to be the design variable in this case. Note that the model always expresses the particular assumptions we made about the relationships among the variables; if the assumptions change, the model must be changed.)		
Building width	BW	[lf] linear feet
Building length	BL	[lf]
Floor area of one floor	FA	[sf] square feet
Number of office modules	NOM	[#]
Total floor area	TFA	[sf] square feet

The "Top-Down" Model

We can now start setting up the model that explains how the performance variable depends on the design and context variables. This is shown by equations, which are numbered so that we can refer to them and later identify the corresponding equations in the spreadsheet. The variables on the right-hand side of each equation which are design variables, are *italicized*; the known context variables are underlined. In order to calculate the performance variable, the last equations all have to have known variables on the right-hand side.

(1) $CCST = TFA * \underline{CPRC}$ [$ = sf*$/sf]

(2) $TFA = FA * NOF$ [sf = sf*#]

(3) $FA = BW * BL$ [sf = lf*lf]

(4) $BW = 2*(ML + 2*\underline{WT}) + CW$ [lf = 2*(lf + 2*lf) + lf]

(5) $BL = 2*\underline{WT} + (NMPF + 4)/2$ [lf = 2*lf + (lf + 4)/2]

(Regardless of the number of floors, the diagram assumes approximately four modules per floor to be set aside for stairs and service areas; the number of modules per floor is assumed to be the guiding design variable, and given the double-loaded corridor, this number is divided by two.)

$$(6)\ \text{NOF} = \text{NOM}/NMPF \qquad [\# = \#/\#]$$
$$(7)\ \text{NOM} = \underline{\text{NLA}}/\text{MAR} \qquad [\# = sf/sf]$$
$$(8)\ \text{MW} = \underline{\text{MAR}}/ML \qquad [lf = sf/lf]$$

We may wish to specify some constraints on certain design variables in this model. For example, because we use a module for the fire stairs, which have a minimum width prescribed by fire codes, for the sake of the usefulness of the module we may limit MW and ML to values equal or greater than 10 feet. Similarly, because there are code restrictions that apply to the minimum corridor width, we may wish only to consider values of CW equal or greater than 6 feet.

Note that the equations here are written down in a sequence starting from the last performance variable we wish to calculate, continuing with increasingly specific equations until all variables on the right-hand side are known. We call this a "top-down" way of writing the model.

The "Bottom-Up" Spreadsheet Model

When we start calculating (e.g., by hand), we must start inserting values of the variables into the last equations

---

Example 1 : Spreadsheet model for office building scheme A (Double-loaded corridor)

Input variables: (Design variable values shown **bold, underlined**)

Name	Symbol	Value	Unit of measurement	Source	Constraint
Net Leasable Area	NLA	50000	sf	Program	
Module area	MAR	500	sf	Program	
Module length	ML	**15**	lf	Design	ML ≥ 10'
Corridor width	CW	**6**	lf	Design	CW ≥ 6'
Wall thickness	WT	**0.8**	lf	Design	
Construction price	CPRC	55	$/sf of TFA	Context	
No. modules per floor	NMPF	**10**	#	Design	

Equations: (numbered as in top-down model)

Variables	Symbol		Value	Unit	Formula
8) Module width	MW	=	33.33	lf	MAR / ML
7) Number of modules	NOM	=	100	sf	NLA / MAR
6) Number of floors	NOF	=	10		NOM / NMPF
5) Building length	BL	=	234.93	lf	2*WT + MW*(NMPF + 4)/2
4) Building width	BW	=	39.2	lf	2 * (ML + 2 * WT) + CW
3) Floor Area (one floor)	FA	=	9209.39	sf	BW * BL
2) Total Floor area	TFA	=	92093.87	sf	FA * NOF
1) Construction cost	CCST	=	$5,065,163		TFA * CPRC

FIGURE A4-1B.   Spreadsheet model for office building example from Figure A4-1A.

(the ones where all the values of the variables on the right side are known): as it were, from the bottom up. This "bottom-up" fashion is also how we must write the model in the spreadsheet. There is an important reason for this rule. When the program starts calculating equations, it will look for values of variables starting at the top and proceeding down. (Each spreadsheet program will specify in the manual whether it proceeds from top to bottom column by column, or from right to left and then down row by row.) If it does not find the needed value in the cells already checked and/or calculated, it will "look ahead" down and to the right. The problem is that the value of that variable may be one from the last time the program has been run, which may or—more likely—may not be what we want. To avoid problems caused by such "forward references," it is best to strictly build the spreadsheet in a sequence that never uses variable values from parts of the sheet yet to be calculated.

Figure A4-1B shows a spreadsheet corresponding to the above model. In addition to the columns used above for variable name, symbol, and unit of measurement, we have a column for the values of the respective variables. This is where we type in the actual values of design variables and context variables. For any intermediate or final performance variable, however, the value will be generated by a formula typed into the cell and identified to the program as a formula. In Excel (the example is an Excel spreadsheet done on the Apple Macintosh computer), a formula is identified by starting the entry with '='.

In the source or formula column, the figure shows a copy of the formula used in the value column. It is shown as text because the initial '=' is omitted. (This practice is highly recommended because it allows one always to check the formula for correctness without having to look at the specific cell in the spreadsheet. In more complex spreadsheets, it is necessary to do this to keep track of what is going on.)

The formula can refer to a variable in two ways: either by identifying the cell by the row and column where we enter the value of that variable, or by name. Any given cell can be given a label or name (again, check the rules for this in the brand of spreadsheet you are using); the names chosen in our model should of course be the mnemonic symbols we introduced above. We could have used the full names of each variable, but that would make the formulae unwieldy and hard to read. Or we could have used letters such as x, y, z. However, in a model with many variables, we would quickly lose track of what each letter stands for; so we use short mnemonic symbols: short enough to make nice, compact equations, and just long enough to remind us of what each symbol stands for.

The figure shows the spreadsheet with one set of design variable values and the corresponding result: a

construction cost of $5,065,163. It also has several additional columns to the right of the formula column. These columns show the results of changing entries in the values for the design variables Module Width (MW) and Number of Modules Per Floor (NMPF). The reader should build this spreadsheet on a computer and explore its use by moving the pointer around and changing values to see what change does to the cost.

The model is valid only for this particular configuration of form and access; changing to a solution with an internal service core, for example, will require a separate model. But, as can be seen, the number of equations needed in each of these "quick and dirty" models is so small that relatively little effort is involved in setting up a set of models for a number of basic solution types. The number of alternative solutions that can then be explored is very large.

Figures A4-2A and 4A-2B show a diagram and a spreadsheet model, respectively, for a different scheme for the same office building—a circular plan.

Figures A4-3A through 4A-3C show different schematic design solutions for a beachfront hotel that can all be calculated with the same model and spreadsheet.

SUMMARY

This brief discussion showed how electronic spreadsheet models of a building scheme can be set up and then used to explore quickly the performance of many variations of the basic scheme. It introduced the recommended practice of first setting up the model in standard "top-down" fashion and then building the spreadsheet in the reverse order in a "bottom-up" sequence.

REFERENCE

A useful book discussing the modeling of problems for spreadsheet analysis is *How to Model It, Problem-Solv-*

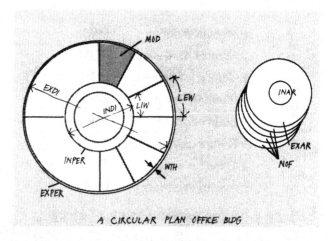

FIGURE A4-2A.   A circular office building plan.

EQUATION #

	VARIABLE	SYMBOL	SOURCE/FORMULA	VALUE	UNIT
	Program net leasable area	PNLA	Program	50,000	sf
	Number of circle divisions	DIV	Decision	10	#
	Length of corridor wall of @ module	LIW	Decision	10	lf
	Wall thickness (ext. wall)	WTH	Design decision	1	lf
13)	Interior perimeter	INPER	LIW * DIV	100	lf
12)	Interior diameter (service area incl. corr)	INDI	INPER/PI	31.83	lf
11)	Interior area incl. corr.	INAR	PI * (INDI/2)^2	795.77	sf
	Module area	MAR	Program	500.00	sf
10)	Exterior area (office space)	EXAR	DIV*MAR	5000.00	sf
9)	Exterior diameter	EXDI	2*((EXAR+INAR)/PI)^0.5	85.90	lf
8)	Exterior perimeter	EXPER	EXDI * PI	269.87	lf
7)	Length ext. module wall	LEW	EXPER/DIV	26.99	lf
6)	Floor area (one floor)	FA	PI*((EXDI+(2*WTH))/2)^2	6068.79	sf
5)	Calculated number of floors	CNOF	PNLA/EXAR	10.00	#
4)	Rounded number of floors	NOF	INT(CNOF) (CNOF rounded)	10	#
3)	Net leasable area	NLA	EXAR*NOF	50000.00	sf
2)	Total Floor Area (Gross)	TFA	FA*NOF	60687.90	sf
	Construction price	BPRC	Context	55	$/sf TFA
1)	Construction cost	BCST	TFA * BPRC	$3,337,834.54	

FIGURE A4-2B.   Spreadsheet model for circular floor plan office building.

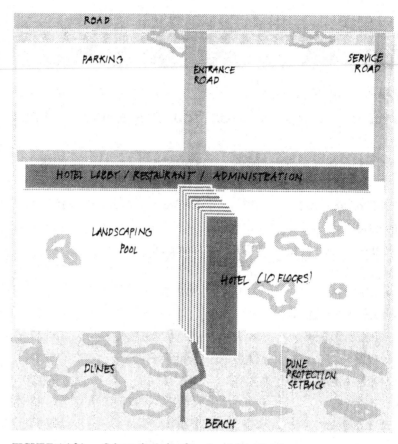

FIGURE A4-3A.   Schematic design for a beachfront hotel.

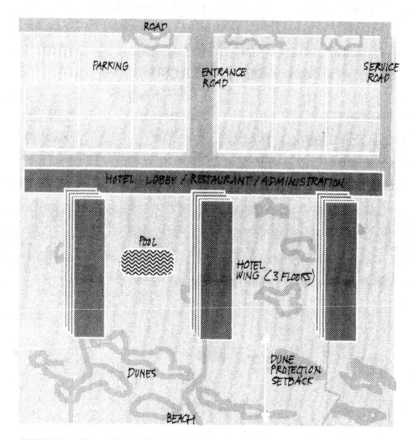

FIGURE A4-3B.   Hotel schematic with three wings.

HOTEL SCHEMATIC #3   COMB - DL CORR

SITE CONTEXT VARIABLES   DESIGN VARIABLES

VARIABLE	SYMBOL	VALUE	UNIT	VARIABLE	SYMBOL	VALUE	UNIT
Site Width	SW	700	f	Room width S/D	RWSD	12.5	f
Site depth	SD	700	f	Room width Suites	RWS	12.5	f
Dune depth	DD	170	f	Room depth S/D	RDSD	24	f
Front setback	FSB	25	f	Room depth suites	RDS	24	f
Side setback	SSB	10	f	Corridor width	CW	6	f
				Parking space width	PSW	9	f
PROGRAM VARIABLES				Parking space depth	PSD	19	f
				Parking lane width	PLW	20	f
Number S/D Rooms	NOSDR	200		Number of floors	NOF	5	#
Number of suites	NOSR	100	#	Access Road width	ARW	30	f
Main Serv. Program	MSPROG	15000	sf	Service Road width	SRW	20	f
Req. no. parking sp.	NPS	300	#	Fire stair 10*25	FSTAR	250	sf
No. bus/RV parking	BRVPS	20	#	Wall thickness	WT	0.5	f
Main serv. efficiency	MSNGR	0.8		Number of comb teeth	NOTH	3	#
Dist. betw. fire stairs	DBFST	200	f	Stair case width	STW	10	f

CALCULATIONS

VARIABLE	SYMBOL	VALUE	UNIT	FORMULA
Parking area length	PARL	630	f	SW-(2*SSB)-ARW-SRW
No. park. spaces per lane	NPSPL	140	#	(PARL/PSW)*2
No. parking lanes req	NOPL	3	#	NPS/NPSPL
Maximum build. length	MXBL	660	f	SW-(2*SSB)-SRW
Front building width	FBW	28.41	f	(MSPROG/MSNGR)/MXBL
Comb length	COLTH	319.11		SD-DD-FSB-FBW-((2*PSD+PLW)*NOPL)-STW
Build. length act.	BLA	4786.58	f	COLTH*NOTH*NOF
Build. length suites	BLDLS	2393.29	f	((2*NOSR)/(NOSDR+2*NOSR))*BLA
No. suites provided	NOSP	96	#	BLDLS/(2*RWS)
Build length for S/D pro	BLDLSD	2393.29	f	BLA-BLDLS
Build.footprint	BFP	49025.66	sf	(MSPROG/MSNGR)+(BLA/NOF)*(RDS+CW+2*WT)+ELAR+NFSTR*250
No. of S/D Rooms	NOSDRP	191	#	(BLDLSD/RWSD)
No. of fire stairs	NFSTR	2	#	COLTH/DBFST
Elevator area	ELAR	200	sf	((((NOF-1)/2)*50)+100)*NOTH
Floor area fl.1	FAGF	30275.66	sf	(BLA*(RDS+CW+2*WT)/NOF)+ELAR+ NFSTR*250
Floor area upper fls	FAU	30275.66	sf	(BLA*(RDS+CW+2*WT)/NOF)+ELAR+ NFSTR*250
Total Floor Area	TFA	170128.32	sf	BFP + (NOF-1)*FAU
Net assignable area	NAA	129877.87	sf	MSPROG+NOSP*(2*RWS*RDS)+NOSDRP*(RWSD*RDSD)
Net to Gross Ratio	NGR	0.76		NAA/TFA
Space betw. buildings	SBBLD	178.33	f	(SW-(NOTH*(2*RDS+CW+2*WT)))/NOTH

BUILDING CONSTRUCTION COST

VARIABLE	SYMBOL	VALUE	UNIT	FORMULA
Main Serv. constr. price	MSPRC	$50.00	$/sf	MSPROG
Hotel constr. base price	HCBPRC	$55.00	$/sf	TFA
Hotel constr. price	HCCPRC	$57.00	$/sf	TF HCBPRC+((NOF-1)*0.5)
Building constr. cost	BCCST	$9,566,064.41	$	(TFA-(MSPROG/MSNGR))*HCCPRC)+((MSPROG/MSNGR)*MSPRC)
Cost per room (unit)	CPR	$33,308.64	$/rm	BCCST/(NOSP+NOSDRP)

FIGURE A4-3C.   Spreadsheet for hotel schematic (shown for three wings).

ing for the Computer Age (McGraw-Hill, 1990) by A. Starfield, K. Smith, and A. Bleloch.

STUDY QUESTIONS

1. (a) Make an educated guess as to which changes in the office building model of Figure A4-1 will bring about the most significant changes in the resulting cost. Identify the two most influential design variables, and change their values upward and downward by 5% (10%). Note the results.

    (b) Identify the three most influential context variables, and then change their values in the same way.

    (c) Propose what you consider the most favorable solution (combination of design variables) for this problem in view of cost. Discuss your answer.

2. Modify the model and spreadsheet for the office building scheme given above to include a single-loaded corridor, with:

    (a) The service spaces on the same side as the office space.

    (b) The service spaces (stairs, restrooms, etc.) on the other side of the corridor.

    Compare the cost for various solutions of this scheme with the results from your experiments with the given model above. Discuss your answer. What would your recommended solution be?

3. Using the same basic assumptions about the program, context variables, and constraints, develop a different design scheme that you think will be more cost-effective, and set up the corresponding "top-down" and "bottom-up" models for it. Explore various variations of this scheme, and propose the solution with the lowest cost. Compare your results with the solutions from exercises 1 and 2; discuss your answer.

Appendix 5
Systems and Models

INTRODUCTION: THE SYSTEMS APPROACH

Throughout the book, recurring references are made to models and modeling. These terms are widely used with differing meanings; so a brief discussion and explanation of their use is in order here.

The connection between the concerns related to economic performance and architectural design can be conveniently discussed using concepts derived from what may be loosely called the "systems approach." Systems science, systems thinking, and the systems approach are all concepts representing a major interdisciplinary effort during the second third of the twentieth century—especially after World War II—to bridge the widening gap in communications between increasingly specialized academic disciplines and sciences. More practical motivations for this effort were those of filling in the missing element in the scientific method—the application element or, we might say, the design element. This was shown most clearly in the emerging discipline of operations research. The systems approach efforts also provided a common conceptual basis for planning, management, political science, research and development, the space program and military planning, and many other areas of application. Cybernetics, the new science of guidance and control systems, and the development of computers were major contributing forces in this movement.

Some of the common principles and explicitly stated or implicit recommendations of this approach are as follows: First, there is the recommendation that whatever is to be studied, researched, planned, managed, and so forth, should be considered as a system. Second, symbolic, mathematical representations—models—

are used as the vehicle for study, simulation, analysis, forecasting, and communication about the work. Third, the solutions and results of the manipulation (simulation) and analysis of the models are used as a basis for recommending what to do in the real-world systems represented by the models.

In the early days of the systems approach, the resulting practical recommendations—supported by computer calculations and simulations that previously would not have been possible—often were quite irresistible. The spectacular successes of the space program, for example, generated a considerable degree of enthusiasm and confidence in these techniques. This led to their overconfident use in areas where the conditions for successful application were not given (e.g., urban and social problems), and equally spectacular failures followed. The terms "system" and "model" (mathematical model) almost simultaneously became very popular and fashionable, and very controversial.

With the development of computers as the tools for such modeling advancing ever more rapidly and the modeling techniques becoming more and more sophisticated, we cannot afford to ignore them; nor can we afford to apply them inappropriately. At the very least, we should know their meaning and some basics about how they should be used.

SYSTEM AND MODEL: DEFINITIONS

Dictionaries define "system" as, for example, "the orderly combination, collection, and arrangement of parts," or "the manner of connection of many parts into a whole." Other views are that the structure (orderliness, organization, connectedness) of our knowledge about something is what we call a "system." In this view, a system is what is inside our minds; we do not know what is "out there" or how it relates to the picture in our heads.

More specific definitions can be found in the General Systems literature; for example, "A system is a set of objects together with relationships between the objects and their attributes" (Hall and Fagen, "Definition of System" in *Yearbook of the Society for General Systems Research* Vol. 1, 1956). This is one of the simplest definitions of system one can find. It is an ateleological definition; that is, it does not speak to whether the system is purposeful or not, or what its purpose might be. As such, it is already controversial: many authors think not only that there are teleological (purposeful, goal-seeking) systems as well as ateleological ones, but that the purposefulness of systems is an essential system feature.

For example, Churchman (in *The Systems Approach*), suggests that in order for us to speak meaningfully of something as a system, it must be teleological; it must have a measure of performance—a "client" whose interests are served by the system in such a way that the higher the measure of performance is, the better these interests are served. Furthermore, it will have the following: teleological components that co-produce the performance measure; an environment (context), defined teleologically or ateleologically, that also co-produces the measure of performance; a decision-maker who uses his or her resources to influence the (partial) measures of performance of the system's components, and hence that of the overall system; and a designer who conceptualizes the nature of the system in such a way that his or her concepts potentially produce actions in the decision-maker that then influence the measure of performance of the system (the designer's intentions are "good" in that he or she seeks to change the system so as to maximize its value to the client). Finally, we assume that the system is "stable" with respect to the designer in the sense that there is a built-in "guarantee" that the designer's intentions ultimately can be realized. Churchman likes to refer to this as the "Guarantor of Decisions."

Whether one adopts this interpretation of a system or not, the measure of performance (as related to the goals and objectives of the client), client, component, context, decision-maker, designer, and so on, are significant elements of systems discussions and analysis.

Models are representations (copies, pictures, images) of something—more specifically, of the systems we study. We speak of models in two major ways. The first is that of a model as simply something copied from reality, such as a scale model of an existing building. The second is that of model as something (in reality or in our imagination) to be copied; for example, a model

student. The second sense can be interpreted as meaning, for example, that "Student x is a good model (copy) of an ideal student," and therefore a good model deserving to be copied in turn. A variation of the second sense is that of a plan (represented, e.g., by means of a scale model or drawings) that deserves to be implemented in reality.

Types of Systems

Systems can be categorized in many ways. For example, we already have discussed the distinction between teleological and ateleological systems. Other distinctions are those between:

- Open and closed systems, depending on whether the forces at work within the system are contained within its boundaries, or originate and terminate outside the system.
- Determinate, indeterminate, probabilistic and stochastic systems, depending on whether the relationships and processes occur in an unchanging, deterministic way or not—whether they occur with a predictable degree of probability, or are characterized by randomness.
- Static and dynamic systems, depending on the extent and pace of change within the system configuration.
- Adaptive, nonadaptive, and homeostatic (equilibrium-seeking) systems, depending on the way the system reacts to its environment—by adapting to it, by returning to an equilibrium state after external disturbances, or by seeking a new equilibrium level.
- "Black box" and "glass box" systems, referring to the state of our knowledge about the internal transformations in the system. If we know and understand the inner relationships, being able to "see" them as if through a transparent lid, we speak of a "glass box"; if we can judge the system's behavior only from its input and output, we consider it a "black box" whose inner workings remain hidden.

System Descriptions

Hall and Fagen's definition of "system," mentioned above, suggests a simple format for describing systems; for any given system, list:

- Its objects (or components).
- The attributes or properties of each component.
- The relationships of the components, and properties.

This can be a useful way to describe virtually anything as a system. In mathematical modeling of systems, a common form of system description identifies the variables of the system (its components), the values of each variable (attributes), and the mathematical functions that describe the relationships among the variables.

A Systems View of Design

Rittel (in "Der Planungsprozess als iterativer Vorgang von Varietätserzeugung und Varietätseinschränkung") offers a teleological systems view of design that expands the above variable–value–function description by distinguishing three types of subsystems of a system that is to be designed:

- The design subsystem and model.
- The context model.
- The performance model.

These subsystems are described, respectively, by design variables, whose values are under the control of the designer, who can set their values; context variables, which are not under the designer's control but must be anticipated or predicted if the designer is to correctly predict the system's behavior; and the performance variables, which measure how well the system serves its purpose and achieves its goals.

The designer's task, in this view, is to correctly identify the relevant variables of the system and to manipulate the design variables in such a way that, given the values of the context variables (which must be collected from the context, estimated, or anticipated), design and context variables together will produce the most favorable values of the performance variables and the best possible value of the overall performance measure.

Understanding design within this frame of reference is tantamount to converting the black boxes of the design, context, and performance subsystems into glass boxes: taking the lid off the boxes, so to speak, to make the inner relationships and processes explicit and visible.

MODELS AS REPRESENTATION OF SYSTEMS

Model types can be distinguished according to the nature of the representation, the nature of the system represented, and the intended use of the model. According to the nature of the representation, the following distinctions are made:

- Iconic models (from the Greek *eikon*—picture, image) visually look like what they represent; for example, architectural scale models. Typically, they are easy to understand but difficult to change.
- Analog models use one kind of reality to represent another. For example, a water supply system with its pipes and flows of liquid can be represented by means of electrical wires and flow of electric current. Models of this kind are more difficult to understand (require more explanation) than iconic models but are used because they are more convenient to make, study, and change.
- Symbolic models use symbols to represent reality, for example, symbols standing for variables that describe the system. They are hard to visualize and therefore most difficult to understand. However, they are the most convenient models to construct, manipulate, and change, and, as mathematical models, very commonly used for analytical purposes. Computer modeling typically uses symbolic models.

Examples of distinctions according to the nature of the system represented are: probabilistic models (representing systems whose behavior follows probabilistic patterns), stochastic models (for systems with random behavior features), and deterministic models (for deterministic systems).

Model purposes or uses lend their names to model types; for example, simulation models (that simulate the behavior of a system over time under different conditions), optimization models (that are used to find optimal solutions for problems), forecasting models, and so on.

All models are means of vicarious inspection—we are looking at the model instead of at the real thing for the general purpose of studying, understanding, and communicating about the thing or system represented.

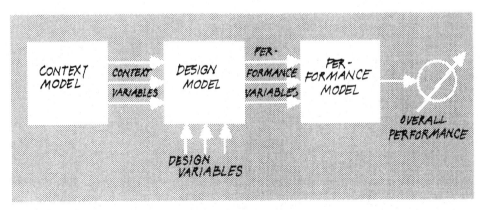

FIGURE A5-1.   A systems view of design (after Rittel).

In the process of doing this, we are making our assumptions and beliefs about the system explicit, and doing this constitutes a significant part of the real value of modeling. It is often said that one of the best ways to understand something is to attempt to construct an analytical model of it; and the only way we can test the validity of assumptions is first to make them explicit.

A special type of model, often used in connection with describing and studying processes consisting of sequences of activities and decisions in time, is the process model, whose visual representation has become familiar as a flow diagram or flow chart. At the extreme, such diagrams are used to specify a systematic procedure in any desired detail. In the case of building economics, we use a rather general (not very systematic) version of such a process representation to describe the building delivery process (see Appendix 2).

STUDY QUESTIONS

1. Practice the ateleological description of selected parts of reality related to buildings; for example, a window, a roof, the circulation system in your design project, and the system of participants in the economic transactions involved in the construction of a building. (Identify system components, their properties or features, and the relationships among them.)
2. Revise the descriptions from question 1 by describing the same phenomena as teleological (purposeful or goal-seeking) systems. You could do this by identifying in these systems the entities suggested by Churchman; in addition to components, their features and relationships identify environment, goal or purpose, measure of performance, client, designer, decision-maker, and guarantor. Or you could describe the systems by listing design, context, and performance variables, and try to describe the relationships among the variables. This can be done first by exploring in a qualitative manner (e.g., "If a increases, b will decrease") and, if applicable, by specifying the mathematical equations that describe the relationships among the variables in a quantitative manner.
3. Examine the systems descriptions of question 2 and explore what happens if you substitute or exchange roles. For example, the client of the economic exchange system of a building project could be the owner (for whom the building is being constructed), or its ultimate user(s), or the various recipients of the money flows: contractors and subcontractors, construction workers, professionals, lending institutions, and so on. The designer could be, other than the architect-designer we tend to think of first, the developer who devises the delivery process; or the owner, or the designer of an industrialized building system, wishing to have the system used for such projects; or the government could be the designer, seeking to develop and put in place viable economic and legal conditions for a building industry in which all participants do well; and so on. Note how the same system changes character, depending on the point of view chosen or on whose purpose it primarily is seen to serve.

Measuring Performance: Evaluation

APPROACHES TO EVALUATION

The problem of determining the economic merit of buildings is embedded in a larger question of the evaluation of alternative design solutions. The language of the systems approach, used for much of the discussion in this book, should not be allowed to obscure the fact that there are several fundamentally different attitudes —with corresponding methodological approaches— toward the question of evaluation in design. Briefly, the four main approaches involve the following:

- Evaluation procedures involving judgments about aspects of the solutions, expressed on some judgment scoring scale.
- The use of monetary scales for all merits, costs, and demerits of alternative solutions, and the use of aggregated monetary measures, ratios, and so forth, formed from such measures (money as the common denominator and value scale).
- "Political" forms of decision-making in which proposals (or aspects of proposals) are subjected to a vote by a group of decision-makers. The aggregated counts and relationships of the votes for and against are then used as the basis for the decision.
- A design solution generated according to the rules of some theory. These rules "guarantee" the validity of the solution so that no alternatives have to be considered.

The following paragraphs discuss these approaches in some detail.

Evaluation Procedures

An example of the first approach to evaluation is the procedure described by Musso and Rittel ("Über das Messen der Güte von Gebäuden"). In this method, sets of evaluation aspects or concerns are identified and organized into a hierarchical tree-like structure of aspects, subaspects, sub-subaspects, and so on; and to the extent possible, objective criteria are used to measure how well a given solution satisfies a given aspect or subaspect. If such criteria are identified, criterion functions (or transformation functions) may be drawn that specify how a person's judgment (on the scoring scale) depends on the values of a given criterion. Weights of relative importance are assigned to the aspects and subaspects. Furthermore, the manner in which partial judgment scores are assembled or aggregated into overall judgment scores must be specified. The competing solution alternatives then are evaluated against the criteria and aspects, resulting in an overall judgment score for each solution; the overall scores then can be compared, and the implied decision rule is that the solution with the best score should be adopted. (Figure A6-1 shows a diagram summarizing the steps for this procedure, which can only be sketched in very rough form here.) The relevant point about economic performance is that with this approach, the economic concern is merely one aspect among others (see Figure A6-2). Its importance in the hierarchy can be expressed by means of the weights of relative importance, and any monetary measure (e.g., cost, returns, profit, etc.) serves as

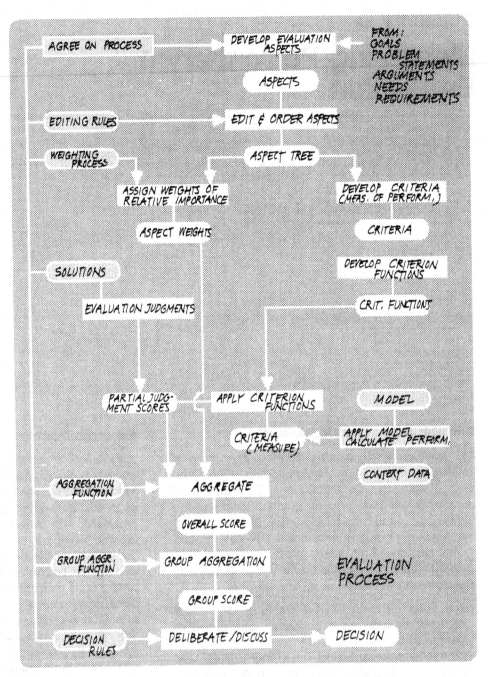

FIGURE A6-1.   A formal evaluation procedure (after Musso and Rittel).

the criterion for that aspect. The common yardstick into which all such measures must be converted is the judgment score (see Figures A6-3A and A6-3B).

Using Monetary Measures as the Common Yardstick: Benefit–Cost Analysis

Another approach is that of using money as the common yardstick. This rests on the assumption that society's monetary system is precisely the common denominator we are looking for, that the monetary prices are indeed the common yardsticks of value by means of which we can conveniently communicate about how we value (evaluate) things, goods, services, buildings, and so on.

Benefit–cost analysis is a widely adopted technique that is based on this view. Indeed, it has been required that proponents of projects to be funded by the government first supply proof of worthwhileness through a Benefit–Cost Analysis. Any solution that is considered

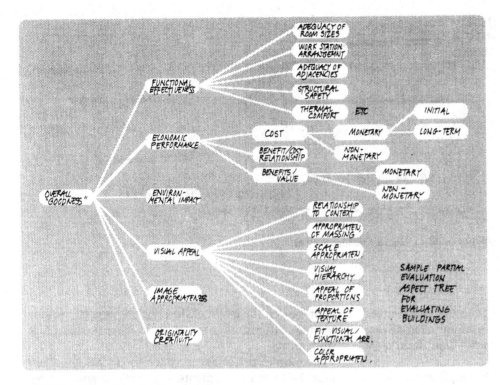

FIGURE A6-2.   An evaluation aspect tree.

must show a Benefit–Cost Ratio well above 1, meaning that the benefits to society are greater than the costs.

The idea underlying Benefit–Cost Analysis is simple and attractive. It assumes that we can adequately express in dollars and cents the value we place on the benefits of a project (even on phenomena not commonly measured in dollar terms), and add this to the actual monetary benefits, if any. We would do the same for monetary and nonmonetary costs, and then form the Benefit–Cost Ratio (BCR) for the proposed project.

If the ratio is at least 1 or better, the project is worthwhile. The benefits are greater than the costs. The higher the ratio, the more desirable it is. (Actually, the government's cutoff value is higher—for example, 1.07. The rationale is that a project should generate at least as much in benefits as the government would earn in interest if the funds for the project were just invested in some "safe" securities instead. The assumed earnings rate for this purpose is called the "discount rate," which reflects the time value of money.)

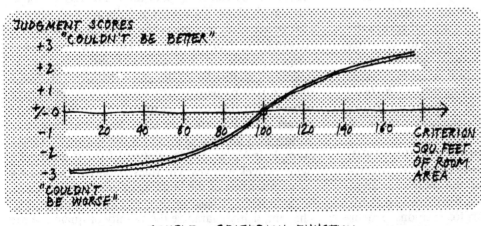

FIGURE A6-3A.   Sample criterion function.

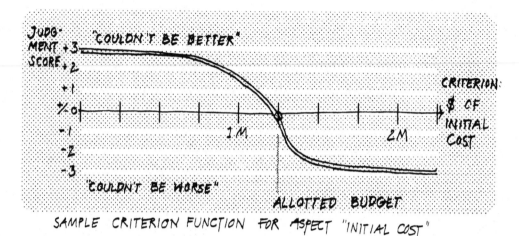

FIGURE A6-3B.   Sample criterion function.

If the ratio is less than 1, the project is undesirable. Among a number of competing solution alternatives, the solution with the highest Benefit–Cost Ratio is to be adopted.

Before adopting this apparently simple and appealing idea too eagerly, we must consider some questions that have been raised about the use of money as a measure of value. Some of these concerns can be answered with appropriate adjustments and manipulations of the measure, whereas others remain unanswered and cause us to have reservations about the validity of the benefit–cost analysis approach that we should keep in mind when using it. (These questions were discussed in Chapter 8.)

Political Forms of Evaluation: Voting

In the political arena, the same problems are traditionally dealt with in a somewhat different fashion. Proposed plans (legislation, government projects) are evaluated not by analyzing their performance against a set of criteria but by raising arguments for and against the proposals, debating their merits or demerits, and then voting. This can take the form of straight yes or no votes, with their number and the ratio of yes to no votes or their distribution in a group being used as the basis for the decision; or the votes can be more discriminating, quantified expressions of preference or confidence in a proposal's worthiness.

Proposals have been made (see Mann, 1977, 1980, 1986, 1990) to establish procedures by which a systematic assessment of the merit of the individual arguments brought forward in such a debate can be transformed into a measure of support for positions in favor of or against a proposal. This approach draws on some of the concepts of the first method above (weighting, goodness/badness judgments, etc.) but leaves the evaluation task in the realm of the political process, without forcing participants to switch to a different conceptual framework.

Economic performance concerns in this approach take the form of arguments, specifically arguments referring to measures or criteria involving monetary amounts or ratios; but these arguments, as in the first approach, are transformed into judgments (votes, or assessment scores for components of arguments) before being made the basis for decision. Figure A6-4 shows a procedure for systematic argument assessment as a basis for decision-making in design and planning.

Rule-Based Design Approaches Eliminating the Need for Evaluation

Strictly speaking, the fourth fundamental approach listed above is not an evaluation approach at all because it avoids the need for formal, comparative assessment or judgment-making in the manner of the other three methods (at least in its pure form). According to this view, which underlies, for example, Christopher Alexander's proposed "Pattern Language" method of designing built environments, but also a number of building systems–based design procedures, the design solution is to be generated by means of a set of components and/or rules (see Alexander et al., *A Pattern Language*). The Pattern Language may serve here as a prototypical modern example for this attitude; a classic example is Vitruvius' description of the rules for designing Greek temples. The rules (patterns) specify how the components of the building or environment are to be combined, at least in their essential configuration. They implicitly guarantee the validity or quality of the solution resulting from their application, and therefore eliminate the need to generate more than one solution. The approach does not rule out variations in detail; in

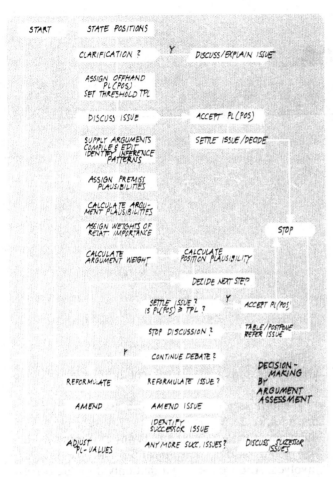

FIGURE A6-4. A systematic argument assessment procedure as the basis for design decisions.

fact, Alexander emphasizes how his patterns can be used "a million times without ever doing it the same way twice." But it is made clear that the variations and details are inessential; the essence of the solution must conform to the patterns. If this is done, the details can be left, presumably, to whimsy, designer judgment, client preferences, or user adaptation.

It is not quite clear how economic aspects are to be dealt with in such approaches. There are two possibilities: either the essential economic viability is assumed to be among the features guaranteed by the use of the rules, and therefore not requiring further attention; or they fall into the category of detail concerns to be addressed after the basic solution, as specified by the rules or patterns, has been determined. Presumably, in this case, they might be dealt with by means of one of the three other approaches described above.

Implications

This brief discussion of various approaches to evaluation in design and design decision-making was necessary to make it clear, first, that economic performance is one of many concerns according to which building proposals are scrutinized, and, second, that the concept of economic performance will play a very different role within each of these approaches.

STUDY QUESTIONS

1. What are the "common yardsticks" or overall measures of performance, into which all other measures are converted, that are used in the four different approaches to evaluation discussed above? Can you think of other measures that might be used as such a common yardstick?
2. Discuss the possible ways in which the issue of economic performance might be treated in rule-based design approaches (in which evaluation is not needed because the rules guarantee the validity of the solution resulting from their application).

Finding the Best Solution: Optimization

INTRODUCTION

Designers sometimes are as quick to promise as their clients are to expect that the "best possible" or "optimal" solution will be found. Such promises are often overconfident to the point of recklessness. Disappointment and estrangement between designer and client are inevitable if the client actually expects more than loose talk and a token effort to look at more than one or two solutions. The following discussion tries to shed some light on the meaning of "optimum," "optimal," "optimization," and related terms, and on how to go about finding optimal solutions if that is called for. The definitions of those terms, the conditions that must be present for claims of optimization to be meaningful, and various approaches to optimization are considered.

It is assumed that the reader is familiar with the concepts of performance, performance variables, constraints, and solution space, as well as with the concepts and procedures discussed in Appendix 6.

THE MEANING OF "OPTIMUM": THE BEST

"Optimum" is a Latin word that means "the best." "Optimal" means "having the property of being the best." "Optimization" means simply the process of finding the best solution.

The concepts seem quite straightforward, but their practical implications are not. What is the "best" solution to a given problem? How is it measured? From our discussion about evaluation, we know that measurement—because it is about good and bad—involves subjective judgments, which means that people will tend to have different opinions about it. The second question then is, how can we achieve agreement about measurement as a condition for making decisions?

We have seen that even the process of explaining to one another the (objective) basis of our subjective judgment about the goodness and badness of proposed design solutions (i.e., objectification) can become very involved. Also we have seen that only if we are persistent and perhaps lucky do we manage to arrive at an acceptably clear connection between our judgment and some "objective," measurable performance variable or criterion. The "criterion function" by means of which we translate values of the performance variable into our judgment score is likely to be different for each of us; we have value systems that can differ in the aspects included in forming judgments, the criterion used, the criterion function, the weights of relative importance assigned to each aspect, and the way in which partial judgments are aggregated into overall judgment scores. Even if we do not remember all of those points, the lesson is that "best" can mean very different things to different people; and if it is so hard even to explain the basis of our judgment to each other, we should be somewhat cautious about assuming that we can make these judgments the basis of common decisions about what constitutes the "best" solution.

CONDITIONS FOR MEANINGFUL CLAIMS OF OPTIMALITY

Let us assume, nevertheless, that those persons involved in a planning or a design situation somehow do agree upon a common definition of what they mean by "good" and "best" in that situation. That is, they have agreed upon some measure of performance by which one can tell which of two solutions is "better" than the other. Then, they can find the value of that performance variable for each solution, compare the values, and decide which solution has the "higher" (or lower, as the case may be) value of that performance measure. Doing this, they will have achieved one of the necessary conditions for talking meaningfully about optimal solutions: there must be an agreed-upon measure (variable) of performance such that each solution will be described by just one value of that variable.

However, this is just the beginning of the problem. Again, what does "optimum" mean? Just the better of the two alternative solutions that were compared? Clearly, that would be too narrow a view, even though we could make it the basis of our understanding; "best" or "optimal" in this case would mean that solution—out of the set of solutions for which we have established the value of a defined and agreed-upon performance measure—with the higher value of that measure of performance.

But if we threw the issue wide open and said "optimal" is to mean the solution with the highest value of the performance measure out of all possible solutions, we would lose our basis of agreement unless we had actually examined all those possible solutions and determined their respective performance. This is the second of the conditions for any meaningful talk about optimization: a well-defined solution space, with well-defined constraints. In other words, the values of the design variables by means of which the solution space is described must be limited within some clear range and not open to infinity. Especially with respect to design problems of the kind that Rittel and Webber (in "Dilemmas in a General Theory of Planning") call "wicked" problems, precisely these conditions are not given; there is no well-defined solution space, no well-defined system of constraints, and so on. So if we persist in calling for optimal solutions, this is meaningful only to the extent that we arbitrarily (and drastically) limit the set of solutions we are able to describe and willing to consider.

There is, finally, another expectation methodologists would like to state with respect to claims of optimality. That is, if a solutions is claimed to be optimal, one should be able to prove that there is indeed no better solution in the solution space.

If anything, these considerations should make us somewhat more cautious in promising "optimal" solutions.

GLOBAL AND LOCAL OPTIMA; SUBOPTIMIZATION

However, if we persist in our (some would say reckless) quest for optimality, we should know about a few other terms and associated pitfalls. For their discussion, it helps to visualize the solution space as a hilly landscape in which the performance measure forms the third dimension, that is, the height or elevation. (See Figure A7-1.) The highest mountain or hill in this landscape is the optimum. If it is indeed the highest in the solution space, it is called the "global" optimum. There are other, smaller mountains; these are called "local" optima. The first potential source of error is that our efforts of optimization would result in finding some mountain, but only a local and not the global optimum. This is, in practice, more difficult to discover than the simple image suggest. (In practice, we do not "see" the landscape.)

Even worse is another possible and very common error, that of suboptimization. Carrying the image of a mountain landscape one step further, the mountain can be conceived of as consisting of layers of judgment scores, just as the rocks of real mountains consist of different layers or strata of stone and soils. The layers correspond to the various subaspects that together make up the overall judgment score or overall performance measure. Because some aspects are more important than others, and also may be easier to measure, the evaluation may be looking at those measures to the exclusion of others, on the unspoken assumption that the highest hill or global optimum also will have the highest invisible layer of the subsequent (or soil type) more or less directly underneath it (see Figure A7-1).

As soon as this assumption is stated, we see how dangerous it is. For the hilly landscape formed by one stratum of rock, or one subaspect, may have its highest

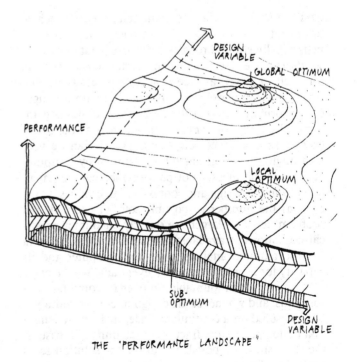

FIGURE A7-1.   The performance landscape.

hills in quite different places than the global optimum. The hills of the subaspect are called "suboptima," and the foolish but widespread practice of fine-tuning solutions according to one or even a few aspects only is called "suboptimization." Because of the importance of the economic aspect in building, combined with the relative ease of measuring costs and even returns, it is fair to say that in building evaluation there is probably more suboptimization going on with regard to economic performance than with all other aspects combined.

THE FUNDAMENTAL APPROACHES TO OPTIMIZATION

Keeping the preceding warnings with regard to uncritical loose talk about optimization in mind, it is interesting to take a quick look at the different ways one can go about finding optimal solutions, provided that the above conditions are satisfied. There are three legitimate fundamental approaches to optimization, as well as one that is not so proper but quite common: the complete enumeration and scrutiny of all solutions in the solution space (then picking the best); the "hill-climbing" approach of successive approximation; the analytical method, using calculus; and finally the approach that strictly speaking has nothing to do with optimization but sounds like it, the approach of successively narrowing down the solution space until there is only one solution left — "the" (only) solution.

Complete Enumeration

Conceptually, the complete enumeration approach to optimization is the least elegant one, as it consists of pedantically going through all possible solutions in the solution space, ascertaining the performance measure for each of them, and comparing the results so that finally one can simply pick the solution with the highest performance. When the number of solutions is intrinsically finite and small enough, such as the set of entries for a competition, this approach can be very meaningful. It is well suited for treatment by computer as long as the measure of performance can be calculated from the design and context variable values. However, as the number of possible solutions to a design problem becomes larger, we quickly encounter practical limitations of available time and effort in carrying out the required analysis. Even the computer and its capacity for mindless rote work eventually will be overcome by a unsurmountable number of combinatorial solutions; and whenever the design or context variables are measured on a continuous scale, such as the length of a building, and any fraction of the units of measurement constitutes a possible solution, the solution set is virtually infinite.

Successive Approximation: "Hill-climbing"

The second approach takes its name, the "hill-climbing procedure," from the hilly landscape visualization of the solution space. It proceeds as a blind man with a stick would go about finding a hill if he were set out on his own somewhere in the landscape. Poking around with the stick, he would feel which direction is "up," take a step or two in that direction, poke around again, and so on, until he had found a spot in which all other directions were "down." This must be the hilltop. Of course, it might not be the global optimum, but only a local one. However, if the process is repeated a sufficient number of times, each time starting at a different spot in the landscape, one may gain a measure of confidence as to the hills identified. The advantage of this procedure is that, compared to the complete enumeration approach, it is significantly less time-consuming. Also, as long as the measure of performance can be calculated, the computer is a most helpful tool in carrying out this kind of analysis. There are many computer optimization procedures on the market that are approximation procedures of one kind or another.

Analytical Approach

This approach is the most elegant and parsimonious one. It uses calculus in describing the performance measure (the landscape surface) as a function of the various design and context variables, finding the first derivative of that function and setting it equal to zero. The solution of this equation describes the spot in the landscape where there is a maximum or a minimum — a hilltop or a valley. The second derivative determines which one is a hill: if the sign of the second derivative is negative, it is a maximum (hill); if positive, it is a valley. If there are several hills, it is necessary to calculate the actual performance values at those spots to determine the highest one.

Sophisticated optimization procedures may use combinations of the hill-climbing and the analytical approaches. For example, in a hill-climbing successive approximation procedure, taking the derivative of the performance function at a given spot (rather than calculating the performance of a set of nearby locations) will immediately give the slope of the hill at that spot, and thus the direction in which to move to reach the top.

Successive Application of Constraints: "The (Only)" Solution

Finally, an approach must be mentioned briefly that was characterized above as slightly less than legitimate, if only to clearly distinguish it from genuine optimization techniques. It should not be called optimization because it does not involve performance measurement as such. All that is done is to introduce successively narrower constraints into the solution space. With skill and luck, this may narrow the solution space until there is only one solution left. This is then convincingly presented to the client as "the" solution or "the only" solution that satisfies the particular set of constraints. It is important to see, however, that it cannot be called optimal or best by any means. There is no measure of performance that would be the basis for the definition of what is good or bad and therefore best; it is defined only relative to the particular set of constraints used, and the sequence in which they are applied. Introducing the constraints in a different sequence might result in a different solution, or even in a situation where the solution space was eliminated altogether, with no solutions left.

OPTIMIZATION IN BUILDING

With the problems discussed above in mind, we would do best not to use the term "optimization" at all when referring to architectural design problems. In most cases, the best we can do would properly be called "suboptimization." This does not mean that we should give up on the idea of working to improve solutions, and using optimization techniques to do so. In building problems, there usually are a number of conditions that simplify this task. For one, the interplay of massing requirements and constraints on various subsystems such as the circulation system, the structural system, natural lighting requirements, and so forth, in practice often reduces the solution space from an infinitely large one to a few meaningful schematic alternatives. Furthermore, there are almost always constraints such as setback requirements, height limitations, density or coverage limits, or combinations of program requirements (e.g., the ratio of required parking spaces to number of apartments, etc.) that further drastically limit the solution space. Finally, the fact that building height in number of floors is a discrete variable (no fractions are accepted as permissible solutions) as well as other similar considerations (the number of office modules or hotel rooms or apartments on a floor always will be an integer) often means that, for practical purposes, the number of possible solutions will be finite and sometimes quite limited. If this is the case, the problem becomes one that can easily be handled by computer, using one of the optimization approaches discussed above—calculus, successive approximation, or even complete enumeration. The performance measure used will likely be confined to one evaluation aspect only, which means that the result will be suboptimization. Spreadsheets are particularly suited for such problems (see Appendix 4), and can be adapted both to

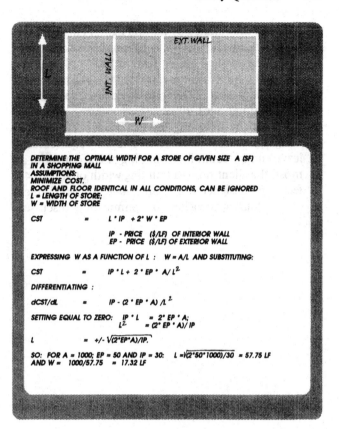

FIGURE A7-2.   A simple optimization example using calculus.

complete enumeration and to successive approximation analysis of a given project.

Figure A7-2 shows a very simple example of an analytical approach to a very simple problem, that of finding the best width-to-length ratio of stores in a shopping mall, where the performance measure is cost only, and where only the context variables of the relative costs of party walls and frontage walls are considered, respectively. As such, it too is a blatant example of suboptimization.

STUDY QUESTIONS

1. (a) If a (one-story) building of 10,000 sf can be designed to have any length, width, and height, with no restrictions on using fractions of feet or inches, is the solution space finite or infinite?
   (b) What would be the answer if the length for the same building is limited to a range of 80 to 200 feet, the width to at least 30 and at most 45 feet, the height to a minimum of 9 feet and a maximum of 12 feet?
   (c) What if the designer chooses to work on a modular grid of $4' \times 4'$ in plan and increments of 2 feet in elevation?

2. Assume that the client is looking for a building that is both reasonably cheap and also well-proportioned

and well-illuminated by natural light. (These are the only concerns at this point.) Would a statement by the architect such as "A 100′ × 100′ building will be best because it is the most economical solution" be considered as referring to an optimal solution or a suboptimal one? Why?

3. Consider the following modifications of the above design problem: While site setback requirements leave enough room for a building of 50-foot width at most, the client needs a building width of at least 47 feet. The designer works on a 4′ × 4′ grid. Which of the four approaches to optimization discussed above best describes the resulting recommendation?

4. Assume that the designer is asked to minimize the cost of the building, which in turn is taken to be a function of the amount of external wall because the roof and floor area will be the same 10,000 sf for all solutions. How would you find the cheapest solution if:

(a) The conditions of question 1b apply?

(b) The conditions of question 1c must be met?

(c) The conditions of question 3 are met?

Try several methods. Again, are you optimizing or suboptimizing?

Appendix 8
Time and the Value of Money

OVERVIEW

One of the most important ways in which the value of money (or of anything else) is altered is by the effect of time. Most people will agree that $100 received now does not have the same value as $100 to be received, say, in five years' time. But what would be equivalent to the $100 received now? And how should the difference be accounted for?

This appendix discusses some conventional concepts and mechanisms for treating this effect of time on value in an orderly fashion. It introduces concepts such as present value, future value, rate of change, annuities, and equivalent uniform annual amounts of money, all relative to the baseline date of an analysis; the activities of discounting and compounding; the phenomena of interest, inflation, escalation, and differential escalation; and appreciation and depreciation. The problem of comparing amount of money at different points in time is seen as one of making appropriate conversions from present to future value, annuity to future or present value, or vice versa. Cash flow diagrams are introduced as tools for visualizing and keeping an overview of the comparison and conversion tasks. The various conversion factors and the corresponding conversion factor tables and their use are discussed.

PRESENT VALUE, FUTURE VALUE; RATES

First some distinctions are needed. If an amount of money, say $100 paid "now" and another amount of money, say $108 paid one year from now, are perceived by someone to be of equal value (in the sense that that person would accept either payment and have no pref-

erence between them), the difference of $8 divided by 100 (8%) is called the rate at which the value changes per year. The amount of $100 paid "now" is the "present value" or "present worth"—more specifically, the present value of the $108 paid in one year's time. The $108 amount is called the "future value" of $100 at year one (i.e., after one year's time), at the rate of 8%. Depending on the area of application or on how the change is achieved, we speak of different rates, as in the following examples:

- The *rate of interest* is to be paid on money borrowed; interest is earned on money lent or invested. In the various processes involving loans, mortgages, deposits of savings, and so on, the differences in value due to the passage of time are expressed by means of the concept of interest. A lender wants to have some benefit or advantage in return for letting the borrower use his or her money temporarily, and states this by specifying some amount that must be added to the loan when it is returned. This amount is called interest. If it is specified as a fixed percentage of the loan, and paid each year, it is called "simple interest." If instead the interest owed is added to the loan, and interest in turn is charged upon the interest, we speak of "compound interest" (see below).

- The *inflation rate* refers to the effect of inflation, the prices of goods and services rising over time. Most of us are painfully aware that most of the things we buy seem to be getting more and more expensive. Another way of expressing the same thing is to say that our money is losing buying power: we are not getting as many apples for a dollar as we did last year. This increase in prices or loss of purchasing power of money is called inflation, and the percentage by which prices rise from year to year is the annual inflation rate. It is sometimes referred to as price escalation, and the corresponding rate as the escalation rate.

- In estimating, and in life cycle cost analysis, attention must be given to the fact that prices do not rise equally fast for all types of materials and labor used in a building project. The difference between the average rate of inflation, which applies to most items estimated, and the higher rate for some items is called the *differential escalation rate*. The opposite effect of prices falling throughout the economy, deflation, is comparatively rare; its effect is described by means of the deflation rate.

- The *depreciation rate* refers to the deterioration or loss of value of something over time; conversely appreciation, the increase in value of something, is described by using the *appreciation rate* (see below). Buildings, as well as other commodities, age and deteriorate, and therefore lose their value over time. This loss of value—due to deterioration or functional or aesthetic obsolescence—is called depreciation, and the rate at which it is estimated to occur is called the depreciation rate.

There are two main uses of depreciation. One is to predict the future value (e.g., the sale value) of a building or other asset; it is oriented to the life span of that asset. The other is for tax purposes; tax laws allow an investor to deduct a depreciation "loss" annually that assumes full depreciation over a specified number of years. (The number of years depends on the building type.) The two uses must be carefully distinguished. For both, there are several accepted methods of estimating depreciation. The simplest is the straight-line method, which assumes that the value is reduced by an equal amount every year for a specified number of years—the life span of the building—either to zero or to some estimated salvage value. For other methods see the tax rules or general engineering economy texts.

- If a building increases in value over time, this increase is called *appreciation*. It is widely assumed that real estate is among the better forms of investment because, among other things, its value (market or sale value) is expected to increase or at a higher rate than regular cost-of-living increases or inflation. The appreciation rate of real estate depends on many things, such as the location, the nature of surrounding development, the general economic conditions, and so on, and is among the most difficult rates to predict over the long run; no general guidelines can be given for it. Sometimes, the building portion of a real estate project is assumed to depreciate or lose value while the

site appreciates. The estimated future value of the project for both site and building then will be determined by the combined effect of building depreciation and site appreciation.

- The *discount rate* in this context is the rate used to determine the present value of some future amount under the assumption that the future amount was earned by investing the present value amount at some average, safe, attractive earnings rate. The term "discount rate" actually is used for two different concepts. One is the process of converting a future amount into its present value equivalent; and there are a number of different philosophies for selecting the actual discount rate in a given situation. The other refers to the interest rate charged by the Federal Reserve Board for money lent to its member banks, which in turn determines the interest rate these banks charge to their prime customers—the "prime rate." The discount rate in this sense is one of the main tools used by the Federal Reserve to influence the economy. Both usages should not be confused with the concept of discount points, which refers to an up-front loan fee charged by mortgage lending institutions in addition to the interest.

The impact of these factors on the long-term economic performance assessment of building is significant. A change of a few percentage points in inflation or interest rates can override the effect of savings of the same percentage magnitude in initial construction costs. Figure A8-1 shows the relative impact of small changes in growth rates. A project of $100,000 initial project cost before financing "now" will have some $6,000 or $7,000 added to the cost if completion is delayed by a year (assuming an annual inflation rate of 6 or 7%), and twice that if the delay is two years. Adding construction loan interest of 11 or 12% on half the project cost (the average outstanding balance of the construction loan) increases the total cost by over 12%, to $112,007 or $112,538, for a construction period of one year and by over 25%, to $125,115 or $126,242, if the period is two years. To offset these increases, the initial net cost would have to be lowered to as little as $78,000—a reduction of over 22%.

COMPARING MONEY AMOUNTS AT DIFFERENT POINTS IN TIME

The problem of comparison of amounts of money at different points in time now becomes that of converting present value amounts into their future value equivalent (at some specified point in time), or vice versa. In order to make a meaningful comparison, the actual date of each future value payment must be given.

THE EFFECT OF INTEREST AND INFLATION ON INITIAL PROJECT COST

MONTHS	INITIAL PROJECT COST BEFORE FINANCING NOW AND W. INFLATION ANNUAL INFLATION: 6.00%	CONSTR. FINANCING INTEREST AMOUNT AT RATE = 11.00% (Applied to 50% OF IPC)	TOTAL	% Increase	INT. RATE 12.00%	TOTAL	% Increase
0 - "NOW"	$100,000.00	$0.00	$100,000.00	0.00	$0.00	$100,000.00	0.00
3	$101,507.51	$1,395.73	$102,903.24	2.90	$1,522.61	$103,030.13	3.03
6	$103,037.75	$2,833.54	$105,871.29	5.87	$3,091.13	$106,128.88	6.13
9	$104,591.06	$4,314.38	$108,905.44	8.91	$4,706.60	$109,297.66	9.30
12	$106,167.78	$5,839.23	$112,007.01	12.01	$6,370.07	$112,537.85	12.54
15	$107,768.27	$7,409.07	$115,177.34	15.18	$8,082.62	$115,850.89	15.85
18	$109,392.89	$9,024.91	$118,417.81	18.42	$9,845.36	$119,238.25	19.24
21	$111,042.01	$10,687.79	$121,729.80	21.73	$11,659.41	$122,701.42	22.70
24	$112,715.98	$12,398.76	$125,114.74	25.11	$13,525.92	$126,241.89	26.24

	7.00%	11.00%			12.00%		
0 - "NOW"	$100,000.00	$0.00	$100,000.00	0.00	$0.00	$100,000.00	0.00
3	$101,760.23	$1,399.20	$103,159.43	3.16	$1,526.40	$103,286.63	3.29
6	$103,551.44	$2,847.66	$106,399.11	6.40	$3,106.54	$106,657.98	6.66
9	$105,374.18	$4,346.69	$109,720.87	9.72	$4,741.84	$110,116.02	10.12
12	$107,229.01	$5,897.60	$113,126.60	13.13	$6,433.74	$113,662.75	13.66
15	$109,116.48	$7,501.76	$116,618.24	16.62	$8,183.74	$117,300.22	17.30
18	$111,037.18	$9,160.57	$120,197.75	20.20	$9,993.35	$121,030.53	21.03
21	$112,991.69	$10,875.45	$123,867.14	23.87	$11,864.13	$124,855.82	24.86
24	$114,980.60	$12,647.87	$127,628.47	27.63	$13,797.67	$128,778.27	28.78

	6.00%	11.00%			12.00%		
0 - "NOW"	$78,000.00	$0.00	$78,000.00	0.00	$0.00	$78,000.00	0.00
3	$79,175.86	$1,088.67	$80,264.53	2.90	$1,187.64	$80,363.50	3.03
6	$80,369.45	$2,210.16	$82,579.61	5.87	$2,411.08	$82,780.53	6.13
9	$81,581.03	$3,365.22	$84,946.24	8.91	$3,671.15	$85,252.17	9.30
12	$82,810.87	$4,554.60	$87,365.47	12.01	$4,968.65	$87,779.52	12.54
15	$84,059.25	$5,779.07	$89,838.33	15.18	$6,304.44	$90,363.70	15.85
18	$85,326.46	$7,039.43	$92,365.89	18.42	$7,679.38	$93,005.84	19.24
21	$86,612.76	$8,336.48	$94,949.24	21.73	$9,094.34	$95,707.10	22.70
24	$87,918.46	$9,671.03	$97,589.49	25.11	$10,550.22	$98,468.68	26.24

	7.00%	11.00%			12.00%		
0 - "NOW"	$78,000.00	$0.00	$78,000.00	0.00	$0.00	$78,000.00	0.00
3	$79,372.98	$1,091.38	$80,464.36	3.16	$1,190.59	$80,563.57	3.29
6	$80,770.12	$2,221.18	$82,991.30	6.40	$2,423.10	$83,193.23	6.66
9	$82,191.86	$3,390.41	$85,582.28	9.72	$3,698.63	$85,890.50	10.12
12	$83,638.63	$4,600.12	$88,238.75	13.13	$5,018.32	$88,656.94	13.66
15	$85,110.86	$5,851.37	$90,962.23	16.62	$6,383.31	$91,494.17	17.30
18	$86,609.00	$7,145.24	$93,754.24	20.20	$7,794.81	$94,403.81	21.03
21	$88,133.52	$8,482.85	$96,616.37	23.87	$9,254.02	$97,387.54	24.86
24	$89,684.87	$9,865.34	$99,550.20	27.63	$10,762.18	$100,447.05	28.78

FIGURE A8-1. The effect of inflation and interest on initial project cost.

Compounding and Discounting

If some amount, say, a deposit in a savings account, increased in value by a given percentage or rate of interest in one year of, say, 10%, the amount in the account at the end of the year would be $110. Leaving both the original amount (which is called the principal) and the interest in the account for another year, both would earn interest: $10 for the principal and $1 for the interest, making a total of $121 at the end of the second year. Continuing this, there would be $133.10 at year three, $146.41 at year four, $161.05 at year five, and so on. Interest is applied to interest upon interest. This process is called "compounding" and the interest rate used in the process "compound interest." Of course, the same process—of finding future value equivalents to the starting amount—can be applied to other types of rates also. Reversing the process, starting from future value amounts and finding the present value equivalent, is called "discounting."

Annuities

Another key concept is that of an "annuity"—a stream of equal payments spread out evenly over a number of periods. Correspondingly, comparison tasks include those of converting annuities into their present value or future equivalents, and vice versa.

Baseline Date

It is customary to use the time when such a comparison (i.e., an economic analysis) is made, as the starting point or baseline date of the analysis period and to refer to amounts of money occurring at that date as "present value" or "present worth." Amounts or payments occurring at all other times are "future value."

Cash Flow Diagram

The task of converting payments occurring at different points in time into their present worth, future worth, or annuity equivalents is made easier through the use of cash flow diagrams. These diagrams consist of a horizontal line representing time and a vertical axis on which payments are entered—cost downward, benefits or payments received upward. They help in maintaining an overview of the pattern of cash flows. It should become a regular habit to draw up the cash flow diagrams for any financial analysis problem involving several payments occurring at different times. Figures A8-2 and A8-3 show typical cash flow diagrams with one-time and recurring payments, as well as cyclical payments, which occur at regular but not necessarily yearly intervals—first in a general form, and then with a building project interpretation.

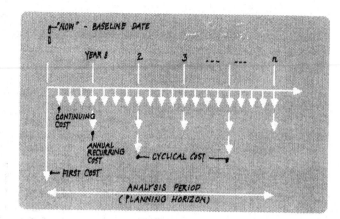

FIGURE A8-2.   Cash flow diagram.

In the conversion, it makes a difference whether a given payment is assumed to occur at the beginning or the end of a payment period (usually years) or in the middle. Figure A8-4 shows two diagrams for the same number of payments, but one with end-of-period payments and the other with beginning-of-period payments.

CONVERSION FACTORS AND TABLES

The main lesson from the preceding discussion was that in order to meaningfully compare any two or more situations (solution alternatives) involving money payments occurring at different points in time, all the payments first must be converted to their equivalent value at some common time. The most frequent approach for this task is the conversion of all future payments into their present value equivalents; but many situations require the conversion from present worth into future worth, or into equivalent annuities, and so forth.

This section presents the formulae for the conversion factors and the corresponding tables of values for

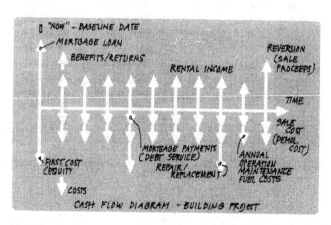

FIGURE A8-3.   Cash flow diagram—building project.

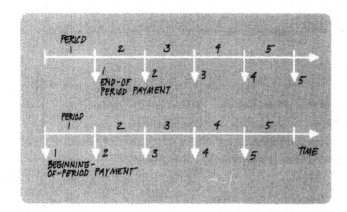

FIGURE A8-4.   End-of-period versus beginning-of-period payment schedule.

the factors, which constitute the main tools for carrying out the conversions.

Conversion of any money amount from its present value into its equivalent future or annuity worth, or from future value or annuity into present worth, and so on, is done with the help of conversion factors. As the following matrix shows, there are six basic factors:

Converting

From: *To>>*

	Present value	Future value	Annuity
Present value	x	1	3
Future value	2	x	5
Annuity	4	6	x

The factors have been given names that differ slightly from author to author. Also, the notation by which the formulae are expressed varies. However, the underlying concepts are the same, and should become familiar to the student of building economics. The following are commonly used notations for the variables occurring in these formulae:

- Present value or present worth: P, PW, PV.
- Future value or future worth: F, FW, FV.
- Annuity: A.
- Interest rate (or any rate of growth to which the formula applies): i. The rate is expressed as a decimal; for example, 12% interest rate is expressed as 0.12.
- Number of periods (years, months, etc.) over which a conversion is to be calculated: n.

The rate and the periods always must correspond. For example, if a formula uses an interest rate, i, that is expressed as an annual rate (for example, interest on a loan to be 10% per year), then the periods in the factor must be in years also. If an annuity such as a mortgage payment is to be calculated as a monthly payment, then the interest rate to be used in the formula must be converted to a monthly interest rate also.

The following is a brief overview of the six factors, each with its corresponding formula and names from a few different sources. The derivations of the formulae are omitted here; they can be studied in any engineering economics or financing textbook, such as Riggs and West's *Engineering Economics*. The formula numbers refer to the numbers in the above matrix.

(1) $(1 + i)^n$ Single-payment future worth factor
    or   Single compound amount factor (SCA)
                  S^n—Future worth of 1
Used to convert a single present-worth amount to future worth at compound interest:

$$FW = PW*(1 + i)^n \qquad (A8\text{-}1)$$

Denoted $(F/P,i,n)$. Future worth, given present worth at i interest over n periods.

(2) $1/(1 + i)^n$ Single-payment present worth factor
                    Single present worth factor (SPW)
                    $V^n = 1/S^n$
                    Present worth of single future payment;
                    Reciprocal of future worth of 1, S^n
Used to convert a single future amount into its present worth equivalent:

$$PW = FW*1/(1 + i)^n \qquad (A8\text{-}2)$$

Denoted $(P/F/i,n)$. Present worth, given future worth at i interest over n periods.

(3) $i(1 + i)^n/((1 + i)^n - 1)$ Capital recovery factor
                    Uniform capital recovery factor (UCR)
                    $i/(1 - V^n)$—Mortgage constant or installment to amortize 1
The capital recovery factor is used to convert a present worth amount to an equivalent series of annuity payments at interest rate i over n periods:

$$A = PW*(i(1 + i)^n)/((1 + i)^n - 1) \qquad (A8\text{-}3)$$

(4) $((1 + i)^n - 1)/i(1 + i)^n$ Series-payment present worth factor
                    Uniform present worth factor (UPW)
                    Present worth of 1 per period
                    or $(1 - V^n)/i$ (sum of present worth factors)

CONVERSION FACTORS

INTEREST RATE i = 8.00%

	1	2	3	4	5	6	7
Periods	Single payment compound amount factor	Single payment present worth factor	Series payment Capital Recovery factor	Series payment present worth factor	Series payment Sinking fund factor	Series payment Compound amount factor	Gradient factor
	$(1+i)^n$	$1/(1+i)^n$	$i(1+i)^n / ((1+i)^n -1)$	$((1+i)^n - 1)/ i(1+i)^n$	$i/((1+i)^n -1)$	$((1+i)^n -1)/ i$	$G((1/i) - (n/((1+i)^n -1))$
n	(F/P,i,n)	(P/F,i,n)	(A/P,i,n)	(P/A,i,n)	A/F,i,n)	(F/A,i,n)	(A/G,i,n)
1	1.08	0.925925926	1.08	0.925925926	1	1	1.06581E-14
2	1.1664	0.85733882	0.560769231	1.783264746	0.480769231	2.08	0.480769231
3	1.259712	0.793832241	0.388033514	2.577096987	0.308033514	3.2464	0.948743223
4	1.36048896	0.735029853	0.301920804	3.31212684	0.221920804	4.506112	1.403959777
5	1.469328077	0.680583197	0.250456455	3.992710037	0.170456455	5.86660096	1.84647159
6	1.586874323	0.630169627	0.216315386	4.622879664	0.136315386	7.335929037	2.276346033
7	1.713824269	0.583490395	0.192072401	5.206370059	0.112072401	8.92280336	2.693664875
8	1.85093021	0.540268885	0.174014761	5.746638944	0.094014761	10.63662763	3.098523941
9	1.999004627	0.500248967	0.160079709	6.246887911	0.080079709	12.48755784	3.491032718
10	2.158924997	0.463193488	0.149029489	6.710081399	0.069029489	14.48656247	3.871313913
11	2.331638997	0.428882859	0.140076342	7.138964258	0.060076342	16.64548746	4.239502955
12	2.518170117	0.397113759	0.132695017	7.536078017	0.052695017	18.97712646	4.595747461
13	2.719623726	0.367697925	0.126521805	7.903775942	0.046521805	21.49529658	4.940206656
14	2.937193624	0.340461041	0.121296853	8.244236983	0.041296853	24.2149203	5.273050755
15	3.172169114	0.315241705	0.116829545	8.559478688	0.036829545	27.15211393	5.594460324
16	3.425942643	0.291890468	0.112976872	8.851369155	0.032976872	30.32428304	5.904625602
17	3.700018055	0.270268951	0.109629431	9.121638107	0.029629431	33.75022569	6.203745807
18	3.996019499	0.250249029	0.106702096	9.371887136	0.026702096	37.45024374	6.492028421
19	4.315701059	0.231712064	0.104127627	9.6035992	0.024127627	41.44626324	6.769688473
20	4.660957144	0.214548207	0.101852209	9.818147407	0.021852209	45.7619643	7.036947794
21	5.033833715	0.198655748	0.09983225	10.01680316	0.01983225	50.42292144	7.29403429
22	5.436540413	0.183940507	0.098032068	10.20074366	0.018032068	55.45675516	7.541181201
23	5.871463646	0.170315284	0.096422169	10.37105895	0.016422169	60.89329557	7.778626369
24	6.341180737	0.157699337	0.094977962	10.52875828	0.014977962	66.76475922	8.006611519
25	6.848475196	0.146017905	0.093678779	10.67477619	0.013678779	73.10593995	8.225381546
26	7.396353212	0.135201764	0.092507127	10.80997795	0.012507127	79.95441515	8.435183825
27	7.988061469	0.125186818	0.091448096	10.93516477	0.011448096	87.35076836	8.63626753
28	8.627106386	0.115913721	0.090488906	11.05107849	0.010488906	95.33882983	8.828882989
29	9.317274897	0.107327519	0.089618535	11.15840601	0.009618535	103.9659362	9.01328105
30	10.06265689	0.099377333	0.088827433	11.25778334	0.008827433	113.2832111	9.18971248

FIGURE A8-5.   Sample conversion factor table.

Used to convert a series of A paid over n periods at i compound interest into their PW equivalent:

$$PW = A*((1 + i)^n - 1)/(i(1 + i)^n) \quad \text{(A8-4)}$$

Denoted $(P/A,i,n)$. P, given A over n periods at i compound interest.

(5) $i/((1 + i)^n - 1)$ Sinking fund factor
    Uniform sinking fund factor (USF)
    Sinking fund factor $i/(S^n - 1)$
    Amount of n periodic deposits that will grow at compound interest i to a specified future amount

Used to convert a future amount into its equivalent uniform series of n periods at compound interest rate i:

$$A = FW*(i/((1 + i)^n - 1)) \quad \text{(A8-5)}$$

Denoted $(A/F,i,n)$. A, given F at i compound interest over n periods.

(6) $((1 + i)^n - 1)/i$ Series-payment future worth factor
    Uniform compound amount factor UCA
    Future worth of 1 per period with interest, or growth at compound interest i of a level series of n periodic deposits: $(S^n - 1)/i$

Used to convert an annuity paid over n periods into its equivalent future worth at the end of the nth period:

$$FW = A*((1 + i)^n - 1)/i \quad \text{(A8-6)}$$

Denoted $(F/A,i,n)$. Future worth, given A for rate i over n periods.

The numerical values of these factors for various interest rates can be found in tables in most engineering economics or real estate textbooks. An example is shown in Figure A8-5.

When doing such calculations by hand or with simple calculators only, it is convenient to use these tables; students should practice in order to be comfortable using the tables to carry out such conversions. When using computers—for example, electronic spreadsheets—it is more practical to use formulae rather than tables to calculate the desired values of the factors.

The table in Figure A8-5 shows a seventh column in addition to the six basic factors discussed above. This is the "gradient factor," used to convert a uniformly increasing series of payments into an equivalent uniform series, which then can be the basis of further conversions as needed. Its formula is:

(7) $$A = G((1/i) - (n/((1 + i)^n - 1))) \quad \text{(A8-7)}$$

where G is the amount by which a base annuity A' increases each period.

Appendix 9
Measures and Indicators of Performance

PERFORMANCE MEASURES AND PERFORMANCE INDICATORS

The problem of assessing the worthwhileness of building projects from an economic point of view is equivalent to that of selecting an appropriate measure of economic performance and then estimating its value for the project under scrutiny. The actual performance of a building can be ascertained only after completion. During the design and planning phase, we need to establish reasonably accurate estimates of the eventual actual performance. Ideally, we should take pains to make such an estimate every time we develop a new, significantly different design solution. Each estimate then guides further design decisions. Genuine economic performance measures for building projects are discussed in Chapters 3 and 5 (costs), 6 (benefits and value), and 7 (measures composed of costs and benefits as well as other measures). This appendix takes a closer look at performance indicators and their role in the design and assessment of buildings. After an overview based largely on the scale of application, a few examples in common use—such as efficiency or net-to-gross (area) ratio, and density—will be discussed in detail.

Look at drawings of a building whose economic performance—say, its initial cost, or its rental return —is important to the owner. Without actually trying to calculate that cost or return, can you tell from the drawings whether it is going to meet the owner's expectations—whether it will be "economical"? Would you know what to change in the design to make it more economical? An experienced designer will claim that yes, one can "see" it. But what can be seen?

Certainly, the drawings do not show cost or rental income.

Experience tells us that if the ratio between the sizes of the areas in the floor plan that can accommodate tenant activities (that are "rentable") and those that serve circulation, storage, restrooms, and other service areas—or, put differently, between the rentable areas and the overall floor area of the building—is high (close to 1), then the building is likely to have an acceptable initial cost and to present a favorable picture of rental income. It is the visual relationship between these areas that can be "seen" by experienced designers and that is used to guide design in an intuitive (but in this case, entirely objectifiable) way. This is the common net-to-gross ratio, or the efficiency ratio.

Used in this way, the efficiency ratio is an example of an indicator of probable economic performance, taking a feature of buildings or plans that by experience is reliably associated with actual performance measures. The use of the net-to-gross ratio is justified by our experience that if a floor plan has a high NGR, it usually will have a low overall cost per square foot of usable or rentable area (given a fixed program of net leasable area), a good relation between initial and annual costs and rental income, and so forth. This allows us to use such performance indicators as shortcut devices to guide the design effort.

It is often too time-consuming to calculate the real performance measures—say, the initial cost of a building, or its rate of return on the owner's investment—during the design process, especially for the many tentative schematic design solutions we like to try out during the early design stages. Also, these measures usually depend considerably on variables outside the designer's control (context variables), so that it is not easy to see how much the resulting estimate is due to our architectural design decisions as opposed to assumptions about these context variables.

Therefore, it is useful to have in-between measures or indicators that can tell the designer how well he or she is doing. They should be easy to calculate and should depend only (or mainly) on the designer's decisions in a straightforward manner. Fortunately, there are a number of such interim indicators of probable performance. They do not in themselves guarantee optimal or even just acceptably good overall performance, but they are helpful and sometimes indispensable in assessing the merit of early design solutions, and in guiding the design. In the example, what the designer is looking at (in an intuitive way, without even having to measure it, though it is of course possible to measure) is the size relationship between the "net" and "total" floor area.

It is important to clearly distinguish these two concepts: A *measure of performance* is defined as a variable that can be measured and that tells the designer of any object or solution for a problem how well the solution serves its purpose or solves the problem. A measure of economic performance for a building is a variable that measures how well the building meets the economic objectives and concerns of the client, owner, or user. A *performance indicator* is a measure that merely indicates (hints at) probable performance as it may be measured by a real performance variable. An indicator describes or measures some feature that is directly or indirectly responsible for the actual, eventual performance of a solution. An indicator of economic performance predicts the level of economic performance; and it does so not in the sense of accurately pinpointing the value of a specific performance measure, but in the sense that if solution A has a "better" performance indicator than solution B, then its performance measure is also likely to be higher than that of solution B. Another way of putting this is that the performance measure and its indicator are strongly correlated: if the latter changes, the other also will consistently change in the same direction. Of course, this correlation depends on the choice of the performance measure itself.

Understanding what makes design solutions more or less economical can be described as understanding the relationship between design variables and such performance indicators. Ideally, we say that we fully understand such a relationship if we can state an equation or model that describes how one variable changes as a result of changes in another variable, and if the variables in the model behave just as the corresponding variables would in reality. The examples discussed below will clarify this explanation as well as the difference between performance measures and indicators.

They are discussed in categories applicable to design at different scales, ranging from urban design to the design of interior spaces in buildings, and construction details.

Performance Indicators in Regulations

There is another important reason for being concerned with performance indicators. Because they are directly linked with design decisions, they often are used in regulations of various kinds, to control development and building design. Such regulations then are expressed as constraints placed either on design variables directly (examples are setback regulations in zoning, or height limits) or on intermediate variables that also serve as performance indicators. (It is rare of find regulations stated in terms of controls on performance variables; most so-called performance regulations do not measure performance variables as such but count "points" based on how many design variable–based constraints are met). Examples are density indicators, such as the floor area ratio or the number of dwelling units per unit of site area.

PERFORMANCE INDICATORS AT DIFFERENT LEVELS OF SCALE—EXAMPLES

Performance Indicators for Urban Design

At the level of urban design, the significant decisions involve the overall development pattern, the street network, zoning, allocation of building types in relation to the street system, open space, and massing considerations. For housing developments, the size of individual sites and choices between different development types, such as detached, semidetached, duplex, row house, walkup, slab block, point block, and so on, are examples of the kind of considerations that would be appropriate at this level.

Most of the performance indicators used at this level have to do with density; and because density is always a ratio made up of two or more variables, there are many density measures, depending on which combination of variables is used.

At the planning level, the concern is usually with population density, expressed as number of people per unit of land (e.g., persons per square mile, persons per acre, households per square mile, households per acre). Also, the number of dwelling units per unit of land often is used.

Economical design at this level often is equated with design for high density: the higher the density, the less the cost per person, dwelling unit, and so forth, for land, utilities, and the urban infrastructure. If infrastructure facilities are limited, the concern becomes one of limiting density of development so as not to overburden the infrastructure. For example, if the existing road system in an area cannot be improved, it can accommodate the traffic generated by only so many housing units. Controls on development then often take the form of density limits.

Site Development

At the level of site development, the access patterns, positioning of buildings relative to site borders, setbacks, and massing are the important decisions. The performance indicators that guide these decisions still are mainly density-related, but the land units in the denominator now do not include the areas of streets and other public amenities. Dwelling units per acre is an important density measure in housing. For housing and other building types, the floor area ratio (FAR), also called the plot ratio or the floor space index (units of floor area per unit of site area), is the most significant measure. Where floor area is less indicative, for example, because floor-to-floor heights are nonstandard, density indicators based on volume may be used instead, such as cubic feet of building volume per square foot of site. Again, the rules of thumb generally agree that the higher the density is (e.g., as measured by the floor area ratio), the more economical the scheme.

Coverage, the ratio of the building footprint to the total site area, and the open space index, the ratio of open (usually "green") space to total site, and the number of units of open space per resident, are examples of other density-related indicators at this level.

Building Geometry

The actual control of indicators such as the floor area ratio quickly switches from urban and site design to the level of design of the building itself. Several design variables become important at this level, one being the choice of access and the internal circulation system — for example, point access, linear access, single- or double-loaded corridor systems. Other indicators are the building height and the building depth, both directly linked to the floor area ratio: the higher and deeper the building, the higher the FAR. However, neither can be increased indefinitely without jeopardizing the usefulness of the floor space created.

The most significant indicator of building schematic economy is efficiency or the net-to-gross ratio (NGR), mentioned above. Depending on the building type, there are some variations of the general definition; it could refer to the ratio of units of Net Leasable Area (NLA) or Net Usable Area (NUA) (also the "assignable" floor area, usually denoted NASF) to Total (gross) Floor Area (TFA). It is a value between 0 and 1.00; the higher it is, the better — that is, the more

efficient the design. For more about floor area and efficiency see Appendix 1.

As the design of the building is finalized, other considerations move into focus. Buildings with simple shapes will cost less to build than complex ones; buildings with a lot of external surface will cost more than simple designs. The perimeter–floor area ratio, formed by the perimeter of a building (typical floor plan) over the enclosed area, is a useful indicator of the building's efficiency or economy in this regard. Energy considerations will lead to a search for solutions with a lower ratio of building surface to volume, if heat loss and heat gain are important concerns. Where the climate permits maintaining comfort by means of natural ventilation and other passive means most of the year, the opposite may be true.

In many applications, the NGR must be considered together with the perimeter–area ratio to achieve the most favorable solution; in certain cases it may be cheaper to add empty floor space (i.e., reducing the NGR) in exchange for achieving a building with less exterior wall area. In such situations the decision must be made by looking at the actual relationship between the costs of the floor area and the exterior wall, not by considering the NGR alone. Increasing the building height (number of floors) beyond a certain point will multiply the area used for stairs and elevators and thus begin to decrease the NGR while still increasing the total floor area and the FAR.

Floor Plan Layout

At the level of the internal organization of unit floor plans, the economy of solutions becomes more difficult to measure. Consumer expectation and regulations call for at least minimum space allocation. One indicator used here is floor area per person. This constrains cost reduction, which obviously is linked to the floor area. Internally, however, dwelling units can be laid out for greater or lower efficiency, which is only inadequately measured by the internal equivalent of the net-to-gross ratio (the actually usable area, excluding circulation areas, over the total area of a unit.)

For some building types, the distances between key elements of the plan are important. Fire codes specify the longest allowable distance between an occupied usable space (e.g., office area) and protected fire exits. For buildings housing functions with a lot of circulation between spaces (e.g., hospitals) the aggregate or the average walking distance between pairs of spaces can be an important indicator of floor plan efficiency, perhaps weighted according to the volume of traffic or frequency of trips, or the nature of the traffic. The distance between emergency care stations and operating theaters in hospitals illustrates the critical importance of this indicator.

Otherwise, much depends at this level on the way an apartment, house, or office is furnished, which is why there is very little in the way of solid general knowledge available on what makes a floor plan costly or less costly. Rules of thumb such as "keeping bathrooms and kitchens close together to minimize the length of wet service runs," as a means of keeping plumbing costs in line, illustrate the kind of guidelines that use the indicator "length of wet service runs," but also show the poverty of useful guidelines at this level.

Construction Detailing

By comparison, once the design has been finalized, the stage of construction document preparation offers numerous opportunities for systematic and reliable cost manipulation. This is so mainly because of the availability of construction cost data manuals and similar data sources, which allow careful review and selection from among a wide variety of materials, building systems, and components. This makes it possible to apply not only value engineering of the structure and construction details but also trade-off analysis in view of initial versus long-term costs—though this is by no means universally done. The increased detail of information at this level means that there are fewer generally useful indicators available to guide decisions, but also that they are less needed because it is easier here to obtain actual cost data. However, some indicators have been used, especially in connection with industrialized building system design.

For example, in building systems that require expensive machinery for the production of each part (and therefore intensive commitment of capital), the number of different types of parts needed for a building often is used as an indicator of the probable economic viability of a system design. In a more general sense, the number of different parts and building components is likely to influence cost in any building and should be watched closely.

In summary, performance indicators can be useful design aids at many different levels of project design, in that they allow the designer to monitor how well the design solution is doing as a direct function of his or her manipulation of design variables. But these tools, too, should be used judiciously, and never in isolation.

STUDY QUESTIONS

1. For a car owner concerned about fuel-economy, which of the following measures is an economic measure of performance, a performance indicator, or neither?

   • Cost of filling a tank of gas.
   • Miles per gallon of gas.

   • Combined cost of gas, maintenance, insurance, and registration per 1000 miles driven.
   • Cost per mile driven.
   • Miles per hour of top speed achievable.
   • Number of miles between fillings.

2. Explore the relationship between cost and development controls. Assume a one-story dwelling unit size of 1000 sf. If the building itself costs $50/sf, its share of the total cost is $50,000. Assume the land cost the developer must charge to the buyer to be $1/sf. Zoning regulations may call for a minimum lot size of, for example, one-half acre, one acre, or two acres. Answer these questions: What is the total cost of the property with building and site in each case? What is the smallest lot size you can get away with if there are not lot size restrictions but only setbacks (e.g., front setback 25 feet, sides and back 10 feet)? What if the cost of the road must be figured in the site cost (e.g., a 16-foot-wide road, costing $10 per square foot)? Half of this must be charged to each lot if the road has lots on both sides. Obviously, this depends on the site dimensions, which in turn depend somewhat on the shape of the house plan. Do you see how this consideration nudges a developer toward narrow and deep lots—with or without lot size restrictions? Assume that the developer is convinced that acceptable floor plans can be designed within a house width of 25 feet. How much would the site and the road add to the cost of the house under these assumptions? How much if you were to make a 30-foot-wide house plan possible (i.e., the lot width would now be $30' + 2*10' = 50'$)? Or a square floor plan? Develop schematic floor plans that would fit these conditions. Each foot of site width gives the architect more freedom to develop the floor plan, but adds $16*8 = 80 of road cost to the cost of the site and therefore the house; and this is before considering the cost of utility lines and other amenities such as sidewalks, drainage ditches, landscaping, and fencing along the street, if such items are needed. Identify the performance measures and indicators applicable in this situation.

3. (a) Explore the relationship between building height, NGR, and initial cost, starting with a two-story building with 100,000 sf of NLA and an area of 15,000 sf (15%) for service areas, evenly distributed over the two floors. The service area (disregarding corridor space) must be repeated for each additional floor. Plot the NGR for alternative solutions of three, four, five, six, etc. floors.

   (b) Develop a simple cost formula (area method)

and plot cost per square foot of TFA and NLA for increasing floors. Before calculating, make an educated guess as to what the curves will look like.

(c) Develop a simple enclosure method cost model for the initial building cost for the purpose of plotting the resulting cost per square foot of NLA and TFA for increasing height. Again, hypothesize what the curves will look like before calculating and plotting.

Appendix 10
Knowing the Future: Forecasting

WHAT WE MUST KNOW
BUT CANNOT CONTROL

Context variables are not under the designer's control, but they determine a systems' performance together with the design variables. To the extent that designers wish to be able to predict or estimate the future performance of their design solutions, they must have knowledge about what values the context variables will or might taken on in the future. This makes the problem of forecasting an intrinsic part of any activity of design and planning for the future.

Any estimate involves some form of assumption about values of context variables that are not known, or not accurately known. There are several possible reasons for this lack of knowledge. Maybe we do not have information that is, at least in principle, available; we just have not had the time to find out how and where to get it, and so on. This is the problem of obtaining data—the values of the pertinent context variables, data known somewhere, but not available to us here and now where we need them. On the other hand, it may be that the things we estimate will only happen at some time in the future, in which case it is not possible to know about them with certainty. The estimate then involves some form of prediction or forecasting of what may happen. The task of forecasting again raises the question of data; any forecast that we wish to be able to explain or objectify turns out to rely on things we already know. We then try to extrapolate, from what is known from the past and the present, into the future. This task involves making assumptions about the specific relationships between the variables involved—developing a model of their behavior. In the attempt,

we may find that the context cannot be predicted with any certainty; any number of possible context situations could occur, and we must find a way to deal with the risks involved in designing for one possible future while another actually is going to happen. We may even find that what will happen in the context depends on our own strategy for design; or on what someone else thinks (predicts) that we will do. Thus, the task of forecasting the future turns into a problem in its own right.

The intertwined issues of data—their sources, reliability, and interpretation—and their use in forecasting and decision-making are the subject of this and the following appendix.

FORECASTING AND
FORECASTING TECHNIQUES

Forecasting or predicting the future always has been one of humankind's great fascinations and concerns. Today, it has become a necessity for governments, industry, and business; and it has grown into a discipline or science in its own right. As architects, we are (or should be) more concerned with events and processes that will occur in the future than most other people are, because of the longevity of our buildings; so we should know some of the principles of this discipline.

Several terms refer to the different attempts to deal with the task of anticipating the future: forecasting, conjecture, prediction, and prophesy; and it is useful to make some basic distinctions among them. They can be ordered, for example, according to how well a person is able to explain the basis of the judgments that go into making a prediction—that is, according to their degree of objectification. (Nothing is said about the reliability or veracity of the predictions resulting from each one.)

- *Forecasting:* This technique is used, for example, in connection with weather forecasts, or estimates of economic activities in the business world. Forecasts usually involve technical, specific, well-described procedures, mathematical models, computer simulations, and the like, based on thorough analysis of past data and experience. We could say that forecasts have in common a high degree of objectified knowledge— that is, knowledge where data, the means by which data were obtained, the model relationships representing how we think the variables are related, and so forth, are explicitly stated.
- *Prediction:* This is more specifically concerned with the if-then mechanisms or laws of nature in the variables involved, but often is seen within a more holistic view of the overall system involved. It too aims at objectifying the forecasting business—so that given the laws and the needed data, anybody could do it—but here judgment based on experience and intuition plays a greater role than in forecasting.
- *Conjecture:* This involves "educated guesses" (from the Latin *conjectura*, thrown together). Although such guesses sometimes can be very intuitive, they still rely on background knowledge of the system and processes at hand.
- *Prophecy:* We might include such phenomena as premonitions and the like in this category. Here, the basis for saying that something will happen cannot be made explicit; prophets cannot say (objectify) how they know what they know, other than in terms of experiences (dreams, ecstatic states of mind, religious experiences) that cannot be shared with others.

For our purposes, we can be concerned only with those techniques that involve some form of objectification.

FORECASTING TECHNIQUES

There are two basic requirements for any systematic, objectified forecasting: first, adequate knowledge about some regularity or "law" according to which things happen (the model) and, second, factual information about the past and current conditions to which that law can be applied (data).

Furthermore, prediction and forecasting must always assume some invariant (nonchanging) aspect, no matter how much the actual concern is with patterns of expected change; the laws involved in the behavior of the system to be forecast are assumed to hold in the future as they have in the past. Any change can be recognized only against the background of features

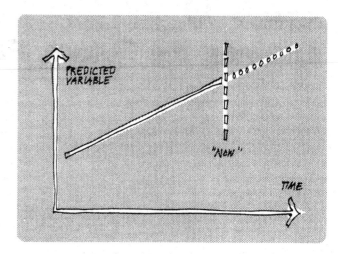

FIGURE A10-1.   Linear trend.

that do not change. The following are some of the most common forms of predictions:

- *Persistence prediction* (the prediction of "no change") is the most basic form of prediction, and an integral part of all other forecasting.
- *Time series extrapolation* is the most common form of forecasting; a trend or trajectory is identified in past observations and then extrapolated into the future. The most critical decision to be made here is that of the form or shape of the assumed trend. Some typical lines and curves that describe trends are:
  (a) *Straight-line or linear functions:* The data are analyzed by means of linear regression techniques: the "least squares method." The straight-line method for calculating depreciation is an example of this type of model.
  (b) *Exponential curves:* Many growth processes in nature can be described by means of exponential functions, at least in their initial stages. The growth of a deposit in a savings account with compound interest and the rise of prices due to inflation are examples of such functions.
  (c) *Logistics functions:* These functions describe growth processes in nature that take place in an environment with constraints. For example, a species in an ecosystem with an unlimited food supply and no predators will grow according to some exponential function as described above. If food is limited, however, the exponential growth will have to slow down and ultimately level off to match the available supply. The characteristic S-shape of such processes can be modeled by a family of functions called logistics functions.
  (d) *Cyclical functions:* Many phenomena in nature or human systems exhibit a pattern of repeated increase and decrease similar to the cycle of the seasons and corresponding variables such as

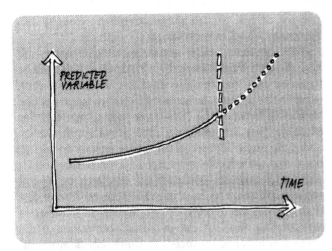

FIGURE A10-2.   Exponential trend.

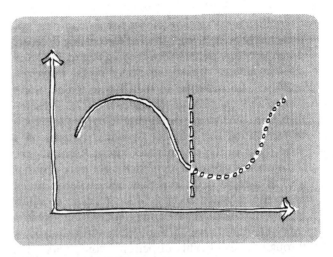

FIGURE A10-4.   Cyclical trend.

the average temperature. Such a rise and fall of the value of a variable often is due to delays built into the system in question, leading to the phenomenon of pairs of variables varying "in phase" or "out of phase," in a way that can be modeled by functions derived from the motion of a wheel —cyclical functions. One well-known example is the "out of phase" movement of meat production and the corresponding meat prices. Because of the natural delay in the reproductive process of cattle, breeding that was increased when prices were high results in an oversupply and falling prices later. Then low prices discourage production so that there are shortages and subsequent rising prices, and the cycle starts again. The effect of such cycles has been countered, for example, by stockpiling of supplies, made possible by the technology of freezer storage. Governments have used purchase, stockpiling, and timed release of such supplies onto the open market to keep prices stable.

These are only a few examples of linear trends, and there are other functions that can be used as the basis for trajectory-extrapolation forecasting models. Applied to the same data, the use of different models can yield very different forecasts; so the choice of the model is one of the most crucial decisions that must be made in forecasting.

Other conceptual approaches to prediction, include the use of:

- *"Social physics"* (attempts to identify the laws that govern social development in much the same way that laws of physics and chemistry govern natural processes).
- *Structural certainties* in the phenomena studied.
- *The overriding problem*, whose treatment will govern what happens in other areas as well.
- *The prime mover* or main driving or motivating force in a society or economy.
- *Scenario-writing* of alternative futures—trying to describe plausible stories of what might happen, then examining the assumptions needed to write these stories to see how likely they are to occur, and in what sequence.
- *Decision theory*, including game theory and simulation gaming, as vehicles for analyzing the behavior of opposing powers or forces, identifying the strategies they will see as the most promising.

Many of these approaches try to address, in some form, the various human motivations that are the driving force behind the change that is predicted. To understand what guides people's behavior, their habits, objectives, desires, dreams, fears, obsessions, and preconceptions must be taken into account.

The difficulty of dealing adequately with human motivations, which are the focus of some of the above approaches, is one of the main criticisms aimed at the following approaches to forecasting:

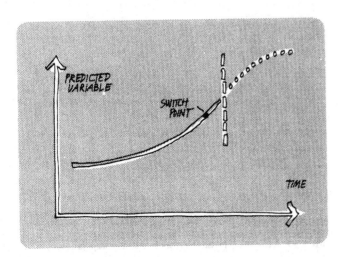

FIGURE A10-3.   Logistics curve.

- *Computerized modeling* of large-scale complex systems has become a major tool of forecasting in recent decades. Much publicity has been given to the "Systems Dynamics" models developed by Jay Forrester, and used, for example, in the Club of Rome report "*The Limits of Growth*" (Meadows 1972). These efforts were the first major attempt to apply these tools to forecasting the behavior of whole systems on a large scale; in the case of the Club of Rome report, the scale was global. Such models have since proliferated at a rapid pace, and they have become much more sophisticated because of advances in computing power (computer size and programming sophistication) and more readily available to individual users through the availability of personal computers.

- An interesting combination of computer models and human decision-making is modern *computerized gaming simulation*: players involved in an urban development scenario make decisions based on their intuitive understanding of the system; these decisions are fed into a computerized model that calculates the system consequences, to which the players then respond with their decisions in the next round, and so on. This makes it possible to "observe" the time-contracted development of large-scale complex systems that depend both on system regularities and on human decisions, over considerable periods of simulated time, and to study the effect of changes in the rule system governing the computer's responses or the human players' decisions. The outcomes of such simulations, especially if repeated many times with different sets of players and context conditions, can be used as forms of forecasts.

- Finally, the *Delphi Method* (Helmer, *Social Technology*) and its derivatives, should be mentioned as a separate approach, consisting of a systematic and organized process of scrutinizing and improving the judgments of a group of experts by successive rounds of exposing the group members to the underlying rationale of the judgments of other members. It first was used to predict the pace of technological developments and their commercial application, but it can be used for all kinds of forecasting scenarios.

PRECAUTIONS IN FORECASTING

In forecasting, as everywhere else, there are numerous possibilities for error and self-deception. If anything, the consequences of errors can be particularly serious here because they are easily compounded over time. It is useful then to look at some precautions that might be taken to avoid the worst pitfalls.

A first principle is to maintain both a critical and a self-critical attitude: critical toward data and toward explanations and assumptions obtained from others, and self-critical toward the assumptions brought into the analysis and interpretation of results.

With respect to data, one must look at where and how the data were collected and recorded. This may be regarded as a concern for statistics, and may require expertise in it, but a measure of common sense applied to reading between the lines often goes a long way. For example, cost data manuals often provide information about the range of projects (in square feet, or dollars, or both) that were used to obtain the average unit prices. It sometimes pays to look at these numbers for several consecutive years to compare the averages. In one cost data manual, the average project size figure was exactly the same as that for the previous year although the average prices had changed. A quick check strengthened the suspicion that the prices had been obtained merely by applying average inflation to those from the preceding year and that no new projects had been surveyed in that category for the updated edition—reason enough to devote some effort to obtaining "second opinion" figures from other sources.

Regarding the "model" part of forecasting, a basic precaution is to consider a number of possible curves that might all explain the data for the past but give significantly different results for the future. Consider a scatter diagram of price data observed over a number of years, say, for the costs of energy used in building operation. (See Figures A10-5A and A10-5B.) What will the price level be in five years' time? If one assumes a straight-line trend, one might predict level A. Being careful, one might look at the possible effects of a 5% or 10% error in the data, which would produce results ranging from A' to A". But what if it is actually an exponential curve about to take off? This might result in a prediction such as B. Or a logistics curve just before its switch point, getting ready to level off again, which would suggest level C? Or yet another type of trend—perhaps a cyclical one? Each with its potential range of

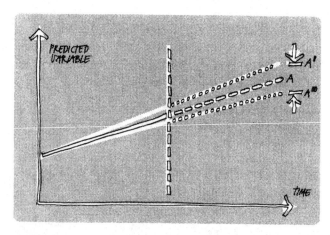

FIGURE A10-5A.   Possible errors in trend extrapolation.

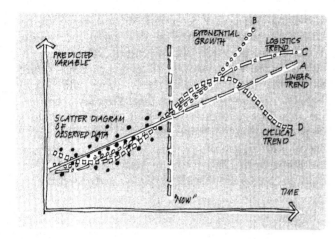

FIGURE A10-5B.   Errors in trend extrapolation.

error compounded into the future? This example should demonstrate the dangers of uncritical acceptance of predictions, no matter how impressively they are backed up by statistical trend analysis. If the wrong type of curve is used, the results not only will be off by some percentage points but will be entirely "out of the ballpark."

Another issue is the relationship between different trends in a larger context. A trend may appear to show signs of consistent exponential growth over a number of years, leading a forecaster to select an exponential model to predict its future development. For example, rental rates for apartments in a particular area may have been rising according to such a trend for some time, suggesting a good market for apartment construction promising high rental income. However, if this trend is seen within the trend of overall income increase for renters in the same area (Figure A10-6), a forecast according to the exponential pattern of the past

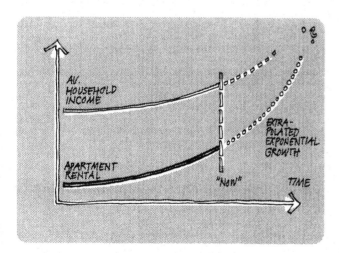

FIGURE A10-6.   Trend in context of other trend.

years is clearly unrealistic—if continued, it soon would exceed the entire household income, which is an impossibility. Unless the income trend is wrong (too pessimistic), the rental rate curve is too optimistic.

How can we guard against such mistakes? There are no panaceas or guarantees. Looking for alternative explanations, inviting and discussing differing opinions with an open mind, double-checking information from one source with data from different sources, probing for any unusual circumstances that may require a different interpretation for the case at hand, are some of the best recommendations available. Although these recommendations sound almost trivial, many serious mistakes made in forecasting contexts for important policy or project decisions can be traced back to failure to heed one or another of these admonitions.

Often, we are so caught up in what we want to see that our interpretations are not only colored by rose-tinted glasses, as it were, but are totally blind to other possible explanations. This situation can lead to our ignoring pieces of data that should be warning signs, or seeing supporting evidence for our plans where there is none. We must constantly remind ourselves to question our own assumptions and motivations. Like the Roman orator who ended every speech, whatever the subject, with the urgent advice that Carthage should be destroyed, we might do well to build constant reminders to "question our assumptions" into all our daily procedures.

How Precise Must Forecasts Be?

In architecture, there usually is no need for great precision of measurements and forecasts, as compared, for example, with the automobile or the computer industry—although architects intrigued with industrialized building systems, prefabrication, and modular coordination have vigorously argued in favor of more precision. Buildings have to accommodate so many different activities that close-fitting to one particular function is impossible, and considerable "slack" must be provided. (This is one of the reasons why the functionalist recipe turned out to be less conclusive in determining the forms of buildings than had been hoped.) Building materials and methods traditionally are not very precise. However, there is one exception: whenever design decisions involve complementary parts, such as nuts that must fit the corresponding bolts, doors and frames, and so forth, the decision on the first part does not require great precision, but the second one does—once the size of bolt has been determined, the nut must fit exactly.

For the purpose of forecasting economic performance, the required precision depends on the closeness of the comparison of alternative solutions: the forecast or estimate must be able to distinguish between the alternatives so that a choice can be made. A rule of thumb for required precision is that the critical performance variable for two alternatives must differ in the last digit to permit a choice to be made.

There is one more aspect to this. So far we have looked mainly at the issue of getting data and using them wisely to make reliable forecasts, but not at the question of what we are going to do with the results in our design and planning. This brings us back to a topic that was postponed for more scrutiny in Chapter 7 on surveying measures of economic performance: decision rules for dealing with situations of risk and uncertainty. What do we do when our forecasting efforts have resulted in the identification of several possible future situations, but only some odds about which will happen—or, worse, not even that? We may even realize that the context is the result of someone else's choice, and that that someone (nature, or some human agent) is looking at our options, trying to guess what we are going to do, and trying to beat us, as in a game. Such problems are discussed briefly in Appendix 11.

Appendix 11
Decisions Under Risk and Uncertainty

TYPES OF DECISION-MAKING SITUATIONS

Most straightforward single measures of performance that are used for making decisions such as selecting the "best" of a set of design alternatives (e.g., the one with the lowest initial cost, or the one that will yield the highest rate of return in *n* years) make tacit assumptions about the context of such decisions that can be crucial. One of these is the assumption that there will be only one future context situation within which the alternatives could exist; another is that it is possible to predict that situation, as well as the expected performance of each of the alternatives in the situation, with certainty.

Starting this assumption explicitly, and perhaps reflecting upon it a bit, is tantamount to recognizing how unrealistic it is in most real-life situations. This is especially true with respect to evaluating the performance of buildings with their long life span, changing users and use patterns, neighborhood, and changing economic conditions.

We see that there will be a number of "things that could happen" to the context of a building that might lead to a set of different situations in which each of the alternative solutions would perform very differently; and more important, we also will have difficulties making judgments about the likelihood of each of the different context situations actually occurring, as well as predicting the performance of the alternative solutions in each of the situations.

An example illustrates this. Your client considers buying a site on a busy street in order to build either a restaurant or a small office building there. Given the nearby mixed-used development, both options would have a reasonable market, and a detailed analysis could probably determine which alternative (restaurant or office) would be more profitable—if conditions remain as they are now. But of course, conditions are not likely to do us that favor. In fact, across the street from the site is another empty site, and rumors have it that is owner is considering selling, and that there are three prospective buyers—a company in need of office space, an oil company that wants to build a gas station, and another restaurant chain. What will be built across the street matters very much in predicting how well our alternatives will do; the unknown future neighbor could provide more customers, or become a competitor, or be somewhat indifferent. How should one analyze such a situation in order to arrive at a prudent decision?

Decision theory and game theory are branches of applied mathematics that have examined such decision-making problems and produced a number of suggestions for how they can be analyzed and resolved.

Three types of decision-making situations are distinguished, which must be treated differently:

1. Decision-making under certainty
2. Decision-making under risk
3. Decision-making under uncertainty

The first situation needs no further comment; it is governed by a simple rule: establish the performance for each alternative and pick the one whose performance is "best" (e.g., lowest cost, or highest return). This is equivalent to the above assumption that things will remain as they are now. But the other two situations need some explanation.

In each case, it is assumed that there are at least two alternative solutions or courses of action. In addition, there are several possible future "states of the world" (or context situations), and our alternative solutions will have different levels of performance for each of them. If it is possible to make reliable estimates about the probability of occurrence of the different states of the world, we are faced with a situation of decision-making under risk. In our example, the states of the world would be: "There will be an office building on the other site," "There will be a gas station," and "There will be a restaurant." Perhaps we might include a fourth: "The other site will remain empty." If we have good reason to believe that all prospective buyers offer the same amount of money and that the owner will flip a coin to determine who gets it, this will be a decision-making situation under risk. We can assign to each of n possible states of the world a probability p_i such that

$$0 \le p_i \le 1.0 \text{ and } \Sigma(p_i) = 1.0$$
$$i = 1, 2, \ldots$$

(The probability p of state i is a number between 0 and 1.0; the sum of probabilities of all n states is 1.0.)

Instead of making such guesses, we could try to figure out which of the three buyers will be able to offer more money for the site, using the approach of income capitalization, estimating an average number of customers for each business and an average profit per customer. For example, the site might support an office building of 50,000 sf. At market rentals, the owner will be able to make a Net-Operating Income (NOI) of $8/sf of NLA per year, that is, $400,000 per year. Using a Net Income Multiplier of 10, this would result in a capitalized value of $4,000,000. If it costs $55/sf of TFA to build the project, with a standard NGR of 0.8, 62,500 sf would be required, and the cost of the project without the site would be $3,437,500; say, $3,500,000. This would leave $500,000 of value for the site, which is the most the prospective developer of the office building could offer. One could make similar calculations for the other businesses to get a feeling for who is most likely to get the site.

When no probability estimates can be made, or, worse, when the situation can be likened to a game being played against an intelligent opponent who selects moves (states) according to a changing strategy in response to his or her anticipation of our own moves, we are in a decision-making situation under uncertainty. For example, the owner of the site in the above example may look at the site of your client, try to anticipate what the client's most likely action will be, and select his or her own strategy accordingly.

To analyze such situations, displays of the alternative choices (moves), the alternative states of the world (opponent's moves), and the performance outcomes (or, in the language of gambling, payoffs) for each combination of alternative solution and state of the world are helpful—the payoff matrix. For the certainty situation, this is a simple array, not yet a matrix (see Figure A11-1). For decision-making under risk, the matrix looks like Figure A11-2. The matrix for the uncertainty situation is the same as the latter except that there are no probabilities.

DECISION-MAKING UNDER RISK: EXPECTATION VALUE

Gambles with known odds are situations of decision-making under risk. Game theory suggests a criterion for this type of decision based on what one would "win" on the average in the long run if one played the same bet many times over. For example, tossing a (fair) coin and betting $1 for each toss for a payoff of $2 if heads turns up, but losing the $1 if tails come up, how much would we win or lose in the long run over many tosses? In this case, we know the answer almost intuitively—we would break even, neither win nor lose. But we might have analyzed the situation as follows: The chance (probability) of heads coming up is 50%. So, over many tosses, half the time we would win $2, with an average gain per toss of $1. From this the average cost per toss must be subtracted, which is the bet of $1 paid whether we win or lose. The result equals zero average gains.

This long-term average payoff is called the "expectation value" of the game, or more precisely, of the

Alternative solutions: >	A1	A2	A3	...	Aj	...	Am
Outcomes >	O1	O2	O3	...	Oj	...	Om

FIGURE A11-1.   Decision-making under certainty.

Alternative solutions: >		A1	A2	A3	A4	Aj	Am
States of the world:	proba-bilities:	O u t c o m e s					
S1	p1	O(1,1)	O(1,2)	O(1,3)	O(1,4)	O(1,j) ...	O(1,m)
S2	p2	O(2,1)	O(2,2)	O(2,3)	O(2,4)	O(2,j)	O(2,m)
...	...	...	...	...	...	...	...
Si	pi	O(i,1)	O(2,i)	O(i,3)	O(i,4)	O(i,j)	O(i,m)
...	...	...	...	...	...	...	...
Sn	pn	O(n,1)	O(n,2)	O(n,3)	O(n,4)	O(n,j)	O(n,m)

FIGURE A11-2.   Decision-making under risk and uncertainty.

choice of betting on the occurrence of heads. It is formally defined as the sum of all possible outcomes (gains or losses), with each outcome multiplied by its particular probability of occurrence, or

$$EV_j = \sum_{i=1,2\ldots}^{n} O_{i,j} * p_i \qquad \text{(A11-1)}$$

where:

EV$_j$ = The expectation value associated with alternative (choice, course of action, solution) j.

O$_{i,j}$ = The payoff or outcome resulting from the choice of alternative j when state of the world i occurs.

p_i = The probability of state of the world i occurring.

If we were in a position to "play" this game many times over, it would obviously be the smart thing to do to select the alternative with the best Expectation Value (e.g., the lowest costs or the highest returns). It is useful to note the distinction between the various considerations that influenced our choice here. We used:

- A straightforward measure of performance or criterion: O. This would be one of the measures of performance discussed in previous chapters.
- A modified criterion based on O and probability of occurrence: EV.
- A decision rule: "maximize the Expectation Value" (i.e., select that alternative that has the best Expectation Value)

These distinction will be very useful when we discuss

ways to deal with uncertainty in the following paragraphs.

Some caution is necessary in using the expectation value to guide decisions. It is appropriate only in those situations where there are indeed repetitive choices. In problems where the decision-maker only has "one shot" at selecting the right solution, it is not prudent to base this selection on the expectation value, especially if the "stakes" (i.e., the costs associated with making the wrong decision, or not achieving the desired payoff) are very high. In gambling, the decision rule has been called the "rich man's strategy" because it may be necessary to endure a long series of losses before achieving the expected large gain. For those who do not have the means to cover such a series of losses, it could be a fatal strategy. Decision theory recognizes some paradoxical features of this criterion, such as the "St. Petersburg paradox": Applying the above decision rule to the determination of stakes in a gamble, it could be modified as follows: The stakes one should be willing to pay for participating in a gamble should be at most equal to the expectation value. Now consider the following proposition: Tossing a coin, you get $1 if tails comes up. If heads comes up, you get to toss again. The payoff for tails is then doubled (as in each subsequent toss); if heads comes up, you get to toss again. How much should you be willing to pay for the privilege of playing this game?

The following are some examples of situations and problems where the expectation value can be a meaningful aid:

- The choice between fixed or flexible (movable, reusable) office partition systems. (Figure A11-9)
- The selection of building components that must be

replaced periodically and involve trade-off considerations between initial and replacement cost and life expectancy (such as light bulbs).

- The determination of insurance rates of various kinds, all of which involve some variation of expectation value.

DECISION-MAKING UNDER UNCERTAINTY

Situations of decision-making under risk can be seen as games against nature—blind, impartial nature which may follow laws of chance (probability) but does not deviate from them. Each single toss of a coin is unpredictable, but in the long run heads and tails will each come up 50% of the time. More significantly, nature does not choose its states of the world, either benevolently or maliciously, in view of our actions or bets. This consistency makes the expectation value a valid decision guide, at least in the long run.

When nature's probabilities are not known, or if the states of the world are the results of actions by other intelligent players who try to anticipate our game moves and select their strategies so as to yield outcomes in their best interest (which usually do not coincide with ours), we are faced with decision-making tasks under uncertainty. Decision theory has developed the following recipes for dealing with such situations, summarized briefly:

- The MaxMin or MinMax rule, recommending that one maximize one's minimum gains or minimize one's maximum possible losses, respectively.
- Bayes' rule of assuming equal likelihood of all possible states of the world and then using the expectation value for that assumption.
- Minimizing one's maximum regret or opportunity costs.
- Establishing a "pessimism/optimism coefficient" (Hurwicz alpha) and selecting the alternative that is best for that coefficient.
- Switching decision criteria and mixing strategies.

The MaxMin or MinMax Rule

This is a pessimistic decision rule, always assuming that the worst situation (state of the world) will happen. Accordingly, the recommendation is to select that course of action for which the worst possible outcome is least undesirable—that is, to "maximize the minimum gains" if the outcomes are defined as gains, and to "minimize the maximum losses" if the outcomes are costs or losses. The payoff matrix of Figure A11-3 describes such a situation. Both players A and B are assumed to choose their course of action knowing the outcomes (payoffs) but not what the other's choice will be. Furthermore, it is a "zero-sum" game: A's gains are B's losses, and vice versa. The payoffs in the matrix are given as A's gains.

The resulting recommendation is for A to select A1 and B to choose B2. Note that the MinMax/MaxMin rule leads to different combinations of actions for the two players, and if they each have reason to believe that the other is also using this decision rule, they may be tempted to change their selection. But the ensuing internal anticipation–switching argument goes on indefinitely.

A special case of this situation is given by the matrix in Figure A11-4. Here, the pair of alternatives A2 and B2 is such that neither player can expect to improve the outcome if he or she deviates from the respective choice. The bold-faced value in the matrix is called a "saddle point"; such points are identified as the lowest in their row but simultaneously the highest in their columns. Obviously, it is useful to analyze a given decision situation for the presence of saddle points.

Bayes' Rule or the Equal-Likelihood Criterion

The eighteenth-century English mathematician Thomas Bayes proposed a decision rule for decision-making under certainty that has become known as Bayes' rule. It recommends that if probabilities are not known, one should decide on the basis of the expecta-

		B's ALTERNATIVES				
		B1	B2	B3	B4	A's MIN
A'S ALTER-NATIVES	A1	-3	0	3	4	-3 MaxMin
	A2	6	-2	1	-4	-4
	A3	0	2	-6	4	-6
	B's MIN:	6	2 (Min Max)	3	4	

FIGURE A11-3.   MinMax/MaxMin payoff matrix.

		B's ALTERNATIVES				
		B1	B2	B3	B4	A's MIN
A'S ALTER-NATIVES	A1	-3	-3	3	4	-3
	A2	6	-2	1	-1	-2 MaxMin
	A3	0	-3	-6	4	-6
	B's MIN:	6	-2 (Min Max)	3	4	

FIGURE A11-4.   MinMax/MaxMin payoff matrix with saddle point.

tion value calculated for the assumption that the states of the world are equally likely. The decision for the first game above then would be settled as shown in Figure A11-5.

Minimizing Regret or Opportunity Cost

When, in a game or in life, we have made a move that does not result in the most favorable outcome, we say that we experience regret—we are sorry we did not pick another move that would have produced a better outcome, given the state of the world or the opponent's moves. A measure of the extent of this regret is the difference between the outcome we actually achieved and the best outcome we could have obtained, given the state of the world or the opponent's move that actually occurred. This measure is also called the "opportunity cost." The corresponding decision rule for uncertainty decisions is to "minimize one's maximum regret," that is, to select that alternative j whose largest regret over the set of i states of the world (or opponent moves) is smallest. Figure A11-6 shows the payoffs in bold, regrets for A italicized, and regrets for B plain, for the game situation introduced earlier, resulting in recommending moves A2 and B3, respectively.

Hurwicz Alpha: Partial Optimism/Pessimism

Few people consider themselves either complete optimists or pure pessimists. Most of us think that we are somewhere in between, in response to each particular case. This means that neither the completely pessimist MinMax/MaxMin decision rule nor its theoretical optimist opposite MaxMax (maximize our maximum gains, not a highly recommended rule) quite represents our true attitudes.

Hurwicz has proposed a "partial pessimism/optimism coefficient," called "Hurwicz alpha," that allows us to express this. The max and min payoffs for each of a player's strategies are plotted on the opposite ends of a graph representing 100% optimism and zero optimism, respectively. Lines are drawn between these extreme payoff points. A decision-maker can now select a point between the extremes that appropriately reflects his or her degree of optimism and then read off which alternative is "best," given that attitude. The graph reveals whether there is one best alternative dominating all others for both MaxMax and MaxMin/MinMax, whether there is a switch point where the line of one alternative intersects that of another, and whether that point is located closer to the optimist or the pessimist side (see Figure A11-7).

			B's ALTERNATIVES				
		probability:	B1 0.25	B2 0.25	B3 0.25	B4 0.25	A's EV:
A's ALTER-NATIVES	A1	0.33	-3	0	3	4	+1 (max)
	A2	0.33	6	-2	1	-4	0.25
	A3	0.33	0	2	-6	4	0
	B's EV: >>>		1	0	-0.67	1.33 (max)	

FIGURE A11-5.   Bayes' rule: the equal-likelihood criterion.

		B's ALTERNATIVES B1	B2	B3	B4	A's maxim. regret
A's ALTER-NATIVES	B's regrets A1 A' regrets	0 -3 9	3 0 2	6 3 0	7 4 0	9
	A2	10 6 0	2 -2 4	5 1 2	0 -4 8	8 (min)
	A3	6 0 6	8 2 0	0 -6 9	10 4 0	9
	B's max regrets:	10	8	6 (min)	10	

FIGURE A11-6.   Regret or opportunity cost.

This kind of analysis has been proposed, for example, to determine which of several alternatives should be used in a strategy of mixed alternatives (see below). Care should be taken to note the use of simplifying assumptions in this approach, such as the straight lines between the best and worst outcomes of a given alternative, as there is no real reason for this. Also, such an extension into a mixed-alternative strategy assumes that the game again is a repetitive one, which may or may not be appropriate to the situation being analyzed.

Mixed-Alternative Strategies

For situations that are truly repetitive, not one-shot decisions, a decision-maker always has the option of selecting one course of action part of the time and switching to other alternatives for the remainder. The choice of alternatives that are combined may in turn rely on several of the above types of analysis and criteria.

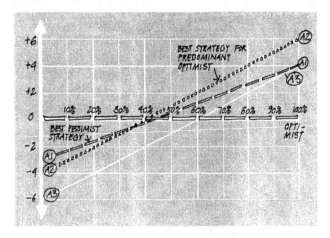

FIGURE A11-7.   Hurwicz alpha — optimism/pessimism coefficient.

UNDERLYING ASSUMPTIONS AND UNRESOLVED PROBLEMS

Decision theory and the theory of games are relatively news as practically applied disciplines (one a few decades old), perhaps were initially overrated, and rely on assumptions that must be clearly understood. For example, the above situations all assume that the payoffs in the matrices can be predicted with reasonable accuracy. Remembering the difficulties we are having just trying to establish a reliable estimate of the initial cost of a building, let alone its long-term performance, this assumption seems somewhat shaky. Furthermore, the situations presented are very simple: two-person, zero-sum games. Reality is always more complex than this; usually there are many more players involved, and the payoffs are not symmetrical as assumed in the zero-sum setup (I win what you lose, and vice versa). Some payoffs may be easily determined, but others are both conceptually nebulous (especially those involving intangibles) and difficult to calculate. Their assessment by different groups affected by the decisions often poses frustrating communication problems, and the alternatives available for selection are never so clearly distinguishable as the examples suggest. In fact, as soon as they are described by an essential design variable measured on a ratio scale, the number of alternatives is infinite. The same holds for context situations (states-of-the-world) or opponent strategies. For example, somebody may consider only two essentially different alternatives, but these may be implemented at any time in the future, which makes for a virtually infinite number of different choices.

The theories assume that decision-makers will behave rationally, with the definition of rationality being something like "trying to maximize one's utility," and that such utility can be adequately measured in the payoffs. Questions such as whether or how the excite-

PRISONER'S DILEMMA		Prisoner B's alternatives	
		Talk	Don't talk
Prisoner A's Alternatives	Talk	7 years / 7 years	10 years / 3 years
	Don't talk	3 years / 10 years	5 years / 5 years

FIGURE A11-8A.   Prisoner's dilemma.

ment of the game itself should be included in these measures usually are carefully excluded at the beginning of theoretical discussions and derivations.

Phenomena such as the St. Petersburg paradox are further reasons for caution. And a final note should be made of a type of decision-making situation that has become famous under the name "prisoner's dilemma" (Figure A11-8A). Its lesson once more is, first, that judgments about things not represented in the payoff matrix do influence decisions significantly, and, second, that in such situations, the pressure to communicate—thus turning the competitive game into a more cooperative one with a different set of rules—becomes almost irresistible. It nicely explains why the corresponding situation in construction, which may be called the "contractor's dilemma" (Figure A11-8B), all too often leads to illegitimate collusion and bid-rigging.

In the prisoner's dilemma, two prisoners are accused of a serious crime that can be proved only if one of or both of them confesses ("talks"), but they will be penalized for lesser offenses if they are uncooperative. The prisoners are faced with the choices and prospect of penalties shown in Figure A11-8A.

DESIGN STRATEGIES FOR POORLY DESCRIBED CONTEXT

Finally, sometimes design decisions have to be made even though we lack adequate data and forecasts. Context conditions may be unknown or may be known only with some degree of approximation because they can be established only in the future, or because finding out would be too difficult, expensive, or time-consuming, or because the nature of the phenomenon itself is unpredictable, random (stochastic), or measurable only as statistical averages and probabilities. Whatever the case, the designer must adopt some coherent attitude toward such situations. Examples of common strategies[1] for designing under such conditions are the following:

- *Designing for qualified extremes.* For example, doors are dimensioned so as to let most people pass through them without bending their heads, but there is always some percentage of taller people who will not be so accommodated. The specific strategy could be expressed in terms of the percentage of the population accommodated as opposed to those not served.

CONTRACTOR'S DILEMMA (Alternatives are high or low bids, payoffs are possible profits in millions of $ if contract is won)		Contractor B's alternatives	
		Bid low	Bid high
Contractor A's Alternatives	Bid low	1 or 0? / 1 or 0?	0 / 1
	Bid high	1 / 0	2 or 0? / 2 or 0?

FIGURE A11-8B.   Contractor's dilemma.

---

[1]*Note:* I owe this list to Professor Horst Rittel's lectures at the College of Environmental Design, University of California, Berkeley.

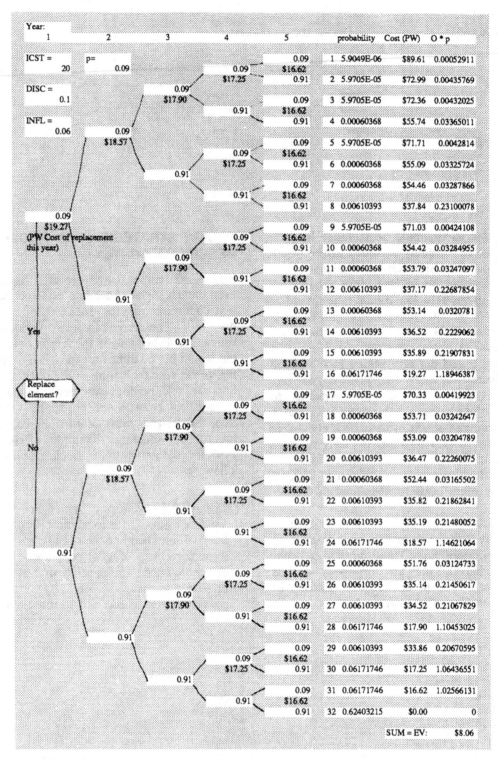

FIGURE A11-9.   Expectation value for office partition problem.

- *Designing for the average.* This clearly would not be a good approach to the door height problem; but door knobs are located at a height corresponding to the average user's hand height. The simplicity of the example obscures how important a conscious look at the chosen strategy can be in less clear-cut circumstances.
- *Gradation,* or designing and providing different solutions for different classes of users. Examples include apartments with different numbers of bedrooms, or shoes that come in different standardized sizes. The strategy can be either for classes of equal increments in size or for classes of equal-size production lots, so that the size increments would be smallest around the midpoint of the population curve.
- *Designing adjustable systems or solutions.* From time to time, such strategies seem to become fashionable, especially among technology-minded students, and then lead to a flurry of inventions for floor plans with movable walls and all kinds of contraptions that have adjustment levers and controls. Good examples are shoes with shoelaces to let the user adjust the tightness, or venetian blinds.
- *Flexible systems.* These take this attitude one step further; the idea is that the solutions should be self-adjusting to the specific condition. The best example is socks where "one size fits all."
- *Fitting the user,* that is, selecting users to match the sizes and requirements of the system in question. This represents the opposite attitude, following the ungracious host of Greek mythology Procrustes with his bed (a guest who was too tall would get a part chopped off; one too short would be—painfully—stretched). This strategy does not work well in free market situations, where users would object. But in certain situations, for example, where weight or special physical attributes are essential (jockeys, astro-nauts), selecting users with those attributes, or training them to conform, may be appropriate.
- *Avoiding design,* that is, rejecting or minimizing designer responsibility. This strategy has been proposed with a number of variations and justifications. "Let the people design, or at least 'participate' in design decisions," for example.
- *Designing for neutrality.* This is another attitude generated by the need to design for unknown user populations and often unknown uses; the standardization of today's office building design, which cannot possibly begin to respond to individual companies' needs, let alone individual users, is testimony to this pervasive strategy. One consequence of this has been the increasing separation of three levels of design: architectural, space planning, and interior design (decoration), justified in theory by the argument that each level can respond to different levels of user concerns, but in reality often resulting in "corporate" interior design solutions as impersonal as the overall architecture into which they are being inserted.
- *Designing for foolproofness.* This strategy assumes the widest possible range of mistakes and misuse by the user, and tries to make the solution immune to the consequences of such behavior.
- *Designing "fail-safe" systems.* This similar strategy considers the possibility that the solution might fail, rather than the user, and aims to design systems that shut off or close down harmlessly if anything goes wrong.

The variety of these strategies in itself suggests that, in any particular situation, the selection of a strategy and its relationship to the data and forecasting techniques available for predicting context and solution performance should be explicitly scrutinized and discussed rather than being taken for granted.

Appendix **12**
Building Value Measures Based on Quality of Occasion and Image

Architects sometimes are frustrated by their clients' preoccupation with the economic "bottom line" aspects of a project and their apparent inability to appreciate the architectural sophistication of a design solution. This frustration often is equally matched by many clients' and developers' disdain for what they see as architects' "head in the clouds" attitude (or even plain ignorance) about those economic necessities. There may be some architects who fit this perception. Their derision of "hack architects" who cater too willingly to mere economic requirements is certainly indicative of a collective sentiment. However, most architects believe that their work does contribute something of genuine value of a project—including economic value, and their frustration stems from difficulty in communicating meaningfully about this with their clients. What is the reason for this communication gap? Is there a way to overcome it?

There are a number of possible reasons for this discrepancy that should be studied. First, the fault indeed could lie with the architect, in that the concerns that guide design have little to do with the economic concerns the client may be primarily worried about. The language used by architects explaining their own work (not always the best guide to understanding), by critics, by theorists, and by professors in design juries often is revealing about what actually guides design.

On the other hand, the fault could lie with the economic concerns. For example, granted that the location and the influence of financing parameters could easily override the effect of an architect's efforts to reduce building cost, does the perceived predominance of these factors lead participants in the process to underestimate the importance of architectural design? Is the

reliance on the net-to-gross ratio as a guide to optimizing a buildings' efficiency overlooking something essential? Are the building financial advisors looking at the wrong measures?

Whatever the answers to these questions, it seems that architects have a problem not only with explaining what their contribution to building value is, but also with measuring this value in order to determine which of two proposed solutions is preferable. Whenever measures of value are unclear, the tendency usually is for more weight to be given to those variables that can be readily measured.

Could it be that the fault lies with both client and architect, for example, in the conceptual frame of reference that is being used to discuss the problem? I suspect that a significant part of the discrepancy has its root here, on both sides.

Once there were great expectations that architecture had a new ability to deal with problems (including economic ones) by the adoption of the concept of "function" as one of its central concerns. This concept held great promise for making useful connections between architectural form and its practical implications, economic and otherwise. Unfortunately, its potential was never fully realized. As the novelty wore off the concept, and some of its shortcomings became visible, architects retreated once again to language that made it difficult to achieve such connections, much less encourage the development of measures of architectural value that could be set against the crude measure of economic feasibility and profitability of the real estate world, whose main link with architecture is the square foot.

But the missed chance of functionalism should not

mean that the search for a better conceptual framework for discussion of both architecture and its economic consequences should be given up altogether. In the following discussion, one attempt to explore an alternate conceptual basis is sketched.

The framework has been described in several papers (see references: Mann 1980b, 1986, 1988). Its central concept is that of occasion — the events or situations that make up human life and that are to be housed in and around buildings. The proposal is made to use this framework both for the discussion of architecture and its purposes and as a basis for deriving measures of building value.

OCCASION, PLACE, IMAGE: TENETS OF THE THEORY

Among the main tenets of the proposed theory are the following: The discussion of architecture can and should be conducted with the concepts *occasion*, *place*, and *image*. Life (in buildings) can be seen as a sequence of occasions. Occasions can be described by specifying the participants and the activities and the functional requirements needed for a place to adequately accommodate them. Participants in an occasion hold images of themselves (who they are, who they ought to be), of their activities (what the activities are like), and of the place accommodating the occasion (what the place is like).

The task of architecture in this view is to design a building (place) that adequately houses an occasion of life in a functional sense, as well as to give the building an image that supports the participants in the occasion —by coinciding with, supporting, or challenging the participants' image of themselves, of their activities, and of the place.

Further key concepts in the theory are *occasion opportunity* and *ease of transition* between occasions. Through the way that they are designed and equipped, places present opportunities for occasions to occur. This is done by making it possible to carry out the activities involved in the occasions, and by making it easy to slip from one activity or occasion into another without having to overcome major physical or psychological barriers. A resulting hypothesis is that places (especially in urban environments) are perceived to be interesting or attractive to the extent that they offer a high number of occasion opportunities together with great ease of transition between them. A popular urban place might be said to have a high "occasion opportunity density." This concept could lend itself to the development of the kind of measures of building value for which we are searching.

APPLICATION TO THE DEVELOPMENT OF MEASURES OF BUILDING VALUE

In order to clarify the connection of this concept to economic measures of value, first consider a somewhat special case: that of occasions that involve the production of items for sale and profit.

If an occasion o is a functional requirement occasion involving the production of some item j for sale in a business, then we can express the value of that occasion per time unit t to the owner of the business as:

$$OCCVAL_o = n*(SALPRC_j - PRODCST_j) \quad \text{(A12-1)}$$

where:

$n =$ the number of items produced per time unit.
$SALPRC_j =$ the sale price of item j.
$PRODCST_j =$ the production cost of item j.

PRODCST includes the cost of input material, labor, energy, services, taxes, insurance, overhead, and so on, for the production of item j. The occasion occurs in a place or a workstation. The value of the workstation (if only one occasion can be carried out there) per time unit is:

$$WSVAL = OCCVAL_o - WSCST_i \quad \text{(A12-2)}$$

where $WSCST_i$ is the cost of the workstation per time unit i — that is, the cost of construction, operation, maintenance, and so on, of the workstation, expressed as annual uniform equivalent of all those initial and future recurring costs.

The workstation value can be understood as the amount the owner should be willing to pay (at most, allowing for profit) for the "rental" of the workstation to carry out the occasion, for one time unit.

The total "purchase" value for a workstation would be the workstation value, capitalized over the expected life span of the workstation or duration of the operation; or the Present Worth equivalent of all expected annual workstation values:

$$TWSVAL = \overset{s}{\underset{i=1,2\ldots}{\Sigma}} WSVAL_i/(1 + DISC)^i \quad \text{(A12-3)}$$

or:

$$TWSVAL = NIM*WSVAL_i \quad \text{(A12-4)}$$

where:

$s =$ the life span of the workstation.
$i =$ the time unit i.

DISC = the discount rate used to convert future to present worth.

NIM = the net income multiplier, used in real estate to capitalize annual net project income into current project (purchase) value.

The value of a space, room, or building then can be seen as the value of all m occasions o than can be carried out in it, minus the cost of the space:

$$SPVAL = \sum_{o}^{m} (TOCCVAL_o) - TSPCST \qquad (A12\text{-}5)$$

where:

TOCCVAL$_o$ = the total capitalized value of occasion o in present value.

TSPCST = the total present value equivalent of all initial and life cycle costs of the space.

m = the number of occasions that can occur in the space.

The above economic production–based interpretation of occasion value and space value can be extrapolated to occasions that are not so directly connected to the economic (monetary) value of produced items. The analogy is based on people's willingness to pay for having spaces built in which no such production occurs; the value attributed to the enjoyment of the space is indicated by the amount people are willing to pay for it. Instead of tying this value to the measure of floor area (square footage), this concept of value is based on the concept of occasion: people are willing to pay for the opportunity to participate in the occasions that will be possible in the space. If a measure for the value of occasions can be established, this will open the way for a new way of measuring—and, more important, predicting (for new projects)—the value of space as a function of the occasions. The value of a space, in this view, would depend on the number and quality of occasion opportunities offered in it. In other words, it depends on the occasion opportunity density of the place.

It is necessary to refine the notion of occasion value somewhat, to be able to express degrees of quality of occasions. The quality of an occasion depends on such concerns as its degree of:

- Adequacy of environmental conditions (light, air, temperature, noise, humidity, view, etc.).
- Appropriateness of the furniture, equipment, and so forth, provided.
- Spaciousness.
- Appropriateness of relationships between the occasions (especially ease of transition between occasions).

- Functionality.
- Appropriateness of image (to a considerable extent).

Some of these factors can be measured relatively easily; others seem to be very difficult if not impossible to measure in any quantitative fashion that then could be related to an economic measure of value. Sometimes it is not necessary actually to measure something in order for awareness of the possibility and the significance of a measure to influence decisions in a beneficial way. On the other hand, measurements that currently seem unattainable actually may turn out to be possible after all, with some additional scrutiny.

To take one unlikely example, how could image appropriateness be measured? What we would look for is something like a coefficient or factor that would reduce the intrinsic measure of value of an occasion to zero if the image of the space in which it was to be accommodated were totally inappropriate, and leave it unchanged at full value if the image were completely compatible.

Adaptations of the "semantic differential" approach to image measurement have been used in environmental design research for such purposes (see, e.g., Bechtel et al., *Methods in Environmental and Behavioral Research*). Based on scaled user responses to a questionnaire with polar opposite characteristics of environments, this approach results in the development of an "image profile" for a given situation. Scaling on, for example, a seven-point judgment scale then permits the derivation of a measure of discrepancy such as the variance between two profiles (e.g., between a profile representing some user's concept of the "ideal" or "desired" space for a given purpose and that user's perceived response profile to an actual space, or a proposed design). The approach might work even better if, instead of the atomistic, detailed descriptors used in the semantic differential, one were to use holistic image concepts (Mann 1988). Either way, one could conceive of a measure based on the degree of discrepancy or congruence between two image response profiles; for example:

$$C_j = 1 - \left(\sum_{k=1,2\ldots}^{p} (w_k * \sqrt{((X_{kd} - X_{ka})^2 / U)}) \right) \qquad (A12\text{-}6)$$

where:

C_j = the coefficient of image congruence for space j.

p = the number of image concepts k (or descriptors in a semantic differential profile).

w_k = the weight of relative importance or significance of image concept k (or descriptor k).

X_{kd} = the score for the desired or ideal image on image concept k (respectively descriptor k).

X_{ka} = the score for actually perceived image on concept (descriptor) k.

U = the range on the scoring scale (e.g., $U = 6$ for a scale $+3$ to -3).

Using such a measure, and assuming for the moment that some occasion value $TOCCVAL_o$ has been established with respect to economic value and the other aspects listed above, the value of a space now can be modified in view of image appropriateness:

$$SPVAL_j = \sum_{o=1,2,\ldots}^{m} (C_{jo} * TOCCVAL_o) - TSPCST \quad \text{(A12-7)}$$

that is, occasion value is modified by the image congruence factor of space j for that occasion.

No actual measurements of occasion value or space value based on these concepts have yet been carried out. But as stated before, merely providing awareness of some more specific things to look for than just square footage may be helpful in designing spaces in buildings and in assessing their value. The important issue here is that these concepts do establish the connection between architectural design, including image and aesthetic factors, and the economics of the building.

Types of Lease Arrangements

The real estate market distinguishes a number of different lease arrangements, according to what the tenant and the owner are paying, respectively. This means that the rental rates quoted for comparable facilities often vary considerably because they are based on different assumptions about what the rent covers. Of course, any kind of analysis aiming at comparative assessment of design alternatives must rest on a basis of "all else being equal"; the assumptions must be the same for all alternatives compared. To ensure that rental rates are comparable, the type of lease arrangement must be specified. The following brief overview will introduce the major conventional types of lease arrangements; in practice, additional variations are possible.

GROSS LEASE

In all kinds of lease arrangements, all the costs involved in owning and operating the building including the owner's profit, if any, ultimately must be paid for somehow. In a lease arrangement where all those costs (real estate taxes, mortgage payments, maintenance, insurance, etc.) except utilities are paid for by the owner, the rent paid by the tenant to the owner must cover all of them; this is a *gross lease*. At the extreme, the rent paid would even cover utilities; this is the case, for example, in hotel or motel room rates, which in addition cover maid service, include use of all furniture, and so on. These kinds of very short-term transactions usually are not seen as lease arrangements, however.

NET, DOUBLE NET, TRIPLE NET LEASE

If the tenant is responsible for paying property taxes in addition to utilities, the rent can be somewhat lower and is called a *net lease*. Making the tenant liable for utilities, property tax, and insurance will result in a *double net lease*, or a *triple net lease* if maintenance also is added to the tenant's responsibilities.

Figure A13-1 shows a diagram of the different types of lease arrangements, according to the cost categories they cover.

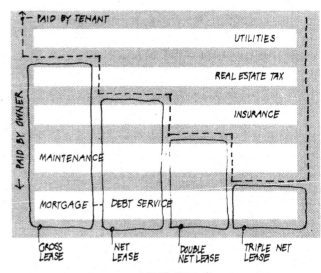

FIGURE A13-1. Types of lease arrangements.

Appendix 14
Glossary

The following definitions include terms that have not necessarily been discussed in this book, but which may be encountered in the literature or in practice in connection with economic performance assessment of real estate projects.

ABD. See *Analysis Base Date.*

Accrued depreciation. The cumulative tax deduction claimed during the holding period, to account for potential loss in utility and thus property value.

Adjustable rate mortgage (ARM). Mortgage arrangement in which the interest rate is not fixed but is periodically adjusted to prevailing market conditions.

Adjusted internal rate of return (AIRR). A version of the internal rate of return that assumes that the positive annual cash flows (after tax) are reinvested at some reinvestment savings rate, and thus contribute to a greater overall accumulation of benefits from the project. See *Modified Internal Rate of Return (MIRR).*

Amortization. The process of paying back a mortgage; the process of an investment returning to the investor, with interest.

Amortization term. The time period between commencement of a loan or mortgage and its maturity date, during which the principal is totally retired.

Amortized mortgage. A mortgage requiring periodic payments, each of which includes partial debt repayment and the interest on the remaining principal balance.

Analog model. A model in which the aspect of reality to be modeled is represented in the model by means of a different kind of reality.

Analysis base date (ABD). The date at which an economic analysis or Life Cycle Cost analysis is done; the date to which all future values are discounted to establish the present value.

Analysis end date (AED). The ending date of the economic performance or Life Cycle Cost analysis period (planning horizon).

Annual debt service. The yearly sum of mortgage interest and principal retirement payments.

Annual value. A uniform annual amount equivalent to the project costs or benefits, taking into account the time value of money throughout the study period (annual worth).

Annually recurring costs. Those costs that are incurred in a regular pattern each year throughout the study period of an economic performance analysis or Life Cycle Cost Analysis.

Annuity. Periodic payments of the same amount at even intervals to some future date.

Area method (of estimating cost). Method of estimating probable construction cost by multiplying the architectural area (total or gross floor area) by an estimated current construction price per area unit (square foot or square meter).

Average outstanding balance. In a construction loan, an assumed percentage of the loan that is treated as having been taken out at the beginning of the construction period and on which the construction loan interest is being paid (usually monthly). This is a simplified assumption for the purpose of convenient approximate estimating of the amount of interest to be paid for such a loan.

Back door financial feasibility analysis (back door approach). Financial feasibility analysis that starts with the rent generated by the project to determine the required initial capital investment and, for example, the construction price at which the project must be built to achieve the assumed performance. If the maximum affordable construction price as established by the analysis is higher than the current estimated market construction price for the same type of project, the project is feasible; otherwise it is not.

Balloon mortgage.   A mortgage in which the monthly payments are calculated as if it were a fixed-rate mortgage, but for which the principal is due in full, for example, after three or five years. It then is refinanced at prevailing interest rates.

Base date; base time (ABD).   The date to which all future and past benefits and costs are converted when a present value method is used for a Life Cycle Cost Analysis. This is usually the beginning of the study period.

Baseline alternative.   Conventional or standard state-of-the-art design alternative (best available) to be included in a Life Cycle Cost Analysis, in response to statutory regulation for energy conservation (together with at least one extraordinary energy-saving design determined to be feasible).

Bayes' rule (equal-likelihood criterion).   A decision rule for selecting the best of a set of alternative courses of action that perform differently under a set of possible future context conditions, when the probabilities of those conditions are not known. The rule recommends that the decision should be made on the basis of the expectation value calculated with the assumption that the probabilities of all alternative futures (context conditions) are equal.

Before tax equity payback (BTEP).   A rate of return formed by the annual before-tax cash flow over the invested equity. Computed cumulatively up to a given year, it expresses how much of the original investment has been paid back up to that year.

Beneficial occupancy date (BOD).   Date of occupancy at which the owner's benefits from the project begin (rent, lease, or the benefit of living/operating in a building).

Benefit–cost analysis.   A method of evaluating projects or investments by comparing the present value equivalent or annual value equivalent of expected benefits to the present value equivalent (or annual value) of expected costs.

Benefit–cost ratio (BCR).   Benefits divided by costs, where both are discounted to a present value or annual value.

Breakeven period.   That period at the end of which the cumulative performances of two alternative projects A and B with different initial costs and annual costs/performance are equal (e.g., A has low initial cost but high maintenance; B has high initial cost but low maintenance). Before that date, alternative A is preferable; afterward, B is better (in this example).

Budget, construction.   The sum established by the owner to be available for construction of the project, including contingencies for bidding and for changes during construction. At each stage of the delivery process, the economic analysis performed at that stage yields a cost estimate that will serve as the budget for the next stage.

Budget, project.   The sum established by the owner to be available for the entire project, including the construction budget, land costs, equipment costs, financing costs, compensation for professional services, costs of owner-furnished goods and services, contingency allowance, and other similar established or estimated costs.

Building footprint.   The area of the site actually covered by the building, usually equivalent to the gross area of the ground floor of the building.

Capital gains basis.   Initial cost minus accrued depreciation.

Capital gains tax.   A tax levied at the time of property resale on property appreciation and straight-line depreciation. If the project sells for more than its depreciated initial cost, the difference is taxed at the capital gains tax rate.

Capital recovery factor.   The factor used to convert a present sum into its equivalent uniform payment series over a specified period (annuities). Most common use: to determine the annual mortgage payments for a loan. $(A/P, i, n)$ or $i(1 + i)^n / ((1 + i)^n - 1)$.

Capitalization.   The process of converting (determining the capital value of) a series of income payments into their present value equivalent.

Capitalization rate (cap rate).   Net operating income divided by the estimated project value. Also: the inverse of the net income multiplier (NIM).

Carrying charges.   Costs incurred by the owner of a project before and during construction that are not directly part of construction costs; for example, site maintenance and security, insurance, real estate taxes.

Cash equity contribution (equity).   The owner's or investor's share (down payment) of cash outlay in a project; that portion of the total initial project cost that is not covered by the loan.

Cash flow.   Generally: stream of payments; the stream of monetary (dollar) values—costs and benefits—resulting from a project investment. Specifically: the before- (Cash Throw-Off, CTO) or after-tax (Spendable Cash After Taxes, SCAT; After tax Cash Flow) benefits generated by a project.

Cash flow diagram.   A diagram that plots income (up) and cost (down) payments against time.

Cash from operations.   Revenue generated minus vacancy loss, expenses, debt service, and income taxes.

Cash on cash rate after taxes (also: *yield,* or *equity yield rate*).   Spendable cash after taxes divided by equity investment.

Cash return on total capital investment.   Sometimes called Rate of Return (ROR) on total capital investment (as opposed to Rate of Return on Equity, ROE); the annual return from a project over its total initial project cost.

Cash throw-off (before tax cash flow).   Revenue minus vacancy and bad debt losses, expenses, and debt service.

Closing costs.   Costs associated with the act of closing (the final settlement between buyer and seller), such as documentary stamp tax, property tax, attorney's fees, recording fees, broker's commission, title insurance, title search fees, and so forth.

Collective goods.   Goods produced or purchased whose use cannot be denied to other members of the community, even those who did not contribute to their purchase.

Component depreciation.   The use of separate depreciation schedules on different elements of a building to account for their loss in utility.

Composite depreciation.   The use of one single depreciation schedule to account for the potential loss of utility in a building (as opposed to component depreciation, where the project's individual components are depreciated separately).

Compound interest.   Interest that accumulates not only on the principal (simple interest) but also on the interest. It results in an exponential growth pattern (e.g., on a savings deposit).

Constant dollars (equals monetary standard for Life Cycle Cost Analysis).   Reflecting the purchasing power of the dollar on the analysis baseline date (ABD), dollars of uniform purchasing power exclusive of general inflation or deflation. Constant dollars are tied to a reference year.

Constraint.   A limit or restriction placed on some variable.

Construction cost (for calculating compensation to the architect).   The total cost or estimated cost to the owner of all elements of the project designed or specified by the architect, including at current market rates (with a reasonable allowance for overhead and profit) the cost of labor and materials furnished by the owner and any equipment that has been designed, specified, selected, or specially provided for by the architect, but not including the compensation of the architect and the architect's consultants, the cost of land, rights-of-way, or other costs that are the responsibility of the owner.

Construction financing.   Generally: the financing process and arrangements of building projects. Specifically: financing arrangements before and during construction (construction loan) where interest rates and other terms including provisions for interest payments are different from those for the long-term or permanent financing (mortgage).

Construction management.   One of several forms of contractual arrangements among the owner, developer, designer, and contractor, aimed at controlling construction costs and time.

Construction period.   For financing cost estimating: the time span between the beginning of construction, when the construction loan has been arranged and interest as well as other financing and carrying charges are beginning to accrue, and the occupancy date, when the construction loan is rolled over into the permanent financing mortgage and benefits from occupancy (e.g., rental income) begin to accrue.

Context variable.   Variables that together with design variables determine the performance of a system but which are not under the control of the designer (must be estimated or obtained as context data).

Contingency; contingency allowance.   A sum included in the project budget (annual expense budget or repair/replacement reserve budget) designated to cover unpredictable or unforeseen items of work or expenses, or changes in the work subsequently required by the owner.

Continuing costs.   As opposed to first cost, costs (e.g., operation and maintenance costs) that will recur year after year or at other intervals.

Contractor's dilemma.   A version of the game theory situation known as prisoner's dilemma that occurs when contractors must decide whether to bid high or low on a project; it demonstrates the strong temptation to communicate outside the rules of the game to solve the dilemma.

Contractor's estimate.   (1) A forecast of construction cost, as opposed to a firm bid, prepared by a contractor for a project or a portion thereof. (2) A term sometimes used to denote a contractor's application or request for a progress payment. As opposed to the architect's/owner's estimate, this is, in the first sense, an estimate of initial costs of a building project that distinguishes between the cost of materials, labor, and overhead and profit, the last being the variables the contractor can manipulate in an effort to produce a competitive bid.

Cost appraisal.   Evaluation or estimate (preferably by a qualified professional appraiser) of the market or other value, cost utility, or other attributes of land or facility.

Cost plus fee agreement.   An agreement under which the contractor (in an owner–contractor agreement) or the architect (in an owner–architect agreement) is reimbursed for the direct and indirect costs of performance of the agreement and, in addition, is paid a fee for services. The fee usually is stated as a stipulated sum or as a percentage of cost.

Costs.   *One-time:* incurred only once in life of project. *Continuous:* incurred periodically throughout the year. *Cyclical:* incurred several times, but not annually. *Annually recurring:* expected to be incurred once a year during the life of project.

Coverage (ratio).   The ratio of the building footprint to the total site area. The inverse of the open space index.

Criterion.   A (usually objectively measurable) variable serving as a performance measure.

Criterion function (transformation function).   A graph or mathematical function describing how someone's evaluation judgment scores (on an agreed-upon scale) depend on a criterion.

Critical path method (CPM).   A network scheduling and control technique for project management.

Critical path scheduling.   Scheduling of (construction) projects using a network scheduling technique (CPM or PERT) that identifies and keeps track of the critical path—the longest sequence of operations/activities and therefore that sequence (path) in which delays in any component activity will result in a corresponding delay in the completion of the entire project.

Current dollars.   Dollars of purchasing power in which actual prices are stated, including inflation or deflation. *Note:* In the absence of inflation or deflation, current dollars equal constant dollars.

Debt cover ratio approach.   Feasibility analysis approach of the "back door" type, starting with the project's rent potential and finding a "revenue-justified" building budget.

Debt service.   The annual or monthly payments to pay back a loan or mortgage, usually such that the payments are of equal size and consist in part of the interest on the loan for that period and in part of a portion of the principal so as to reduce the mortgage balance.

Default ratio (breakeven point).   Operating expenses plus debt service over potential gross income. This ratio determines the point at which a further increase in expenses or reduction in income would cause the property to start losing money.

Deflation (opposite of inflation).   A decline of overall consumer prices, usually accompanied or caused by a short supply of money.

Depreciation.   The phenomenon of loss of utility and/or value of, for example, a building over time; the process of calculating the resulting remaining value at any given time; the activity of calculating the loss of value during one period (e.g., a year).

Depreciation rate.   The annual rate at which the property is estimated to lose utility and thus value.

Design-build.   An approach to construction project management in which the architect and the builder form a team from the beginning of the project as one firm and together guarantee a price at an early stage; a means of eliminating bidding with its potential surprises in project cost after completion of the construction drawings.

Design variable.   A variable that is under the full control of the designer.

Differential escalation rate; differential price escalation rate.   The expected percent difference between the rate of increase assumed for a given item of cost (such as energy) and the general rate of inflation.

Discount factor.   A multiplicative number (calculated from a discount formula for a given discount rate and interest period) that is used to convert costs and benefits occurring at different times to a common time.

Discount points.   A loan fee that is charged by the lender at the closing of a mortgage or a loan, expressed as percentage points of the full loan amount.

Discount rate.   (1) The rate of interest reflecting the investor's time value of money, used to determine discount factors for converting benefits and costs occurring at different times to a base time. *Note:* The discount rate may be expressed as nominal or real. (2) The rate used to convert any future amount or annuity to its present worth equivalent. (3) The interest rate charged by the Federal Reserve Board to the banks borrowing money from it.

Discounted payback period (DPP).   Number of years required to recoup an investment through net savings it provides, with the true value of money and cost escalation taken into account. It is that period, measured from occupancy, which if used as analysis period for LCC would result in net PW savings of zero; or DPP equals that number of years measured from the date of occupancy which, if used as an analysis period for LCCA, would result in a SIR (savings–investment ratio) of 1.0.

Discounting.   A technique for converting cash flows that occur over time to equivalent amounts at a common time. Application of the concept of time preference for money, by which benefits received in the future are valued less than those received in the present.

Distributable cash after taxes.   Cash from operations plus

the working capital loan. This is the final indication of project performance.

Double net lease.   A lease arrangement in which the rental rate or lease is calculated to cover debt service and maintenance plus the owner's overhead and profit, and the tenant is responsible for utilities, real estate taxes, and insurance.

Draw schedule.   The schedule of payments as needed to pay contractors and subcontractors for completed parts of construction work.

Economic evaluation methods.   A set of economic analysis techniques that consider all relevant costs associated with a project investment during its study period, including such techniques as life-cycle cost, benefit-to-cost ratio, savings-to-investment ratio, internal rate of return, and net savings.

Economic life.   That period of time over which an investment is considered to be the least-cost alternative for meeting a particular objective.

Effective income.   Income, usually rental income, obtained from a building, after accounting for losses due to vacancy and bad debts.

Efficiency ratio (net-to-gross ratio).   The ratio of net usable (leasable, rentable) floor area over the total or gross floor area.

Enclosure method.   A method for estimating initial cost of buildings that considers the enclosure components of a building separately (e.g., roof area, exterior wall area, etc.).

Equity (cash equity contribution).   Total project cost minus mortgage amount; the amount of money an owner of a construction must contribute out of his or her own funds; the amount of money an owner has invested in a project (even through mortgage payments).

Equity buildup.   Increase in the investor's share of the total property value resulting from mortgage principal retirement and property appreciation.

Equity capitalization rate.   Sum of before-tax cash flows divided by the original equity investment.

Equity discount rate.   An annual percentage rate, which is used to determine the present value of the project cash flow and reversion.

Equity payback (after tax).   Sum of after-tax cash flows divided by the cash equity contribution.

Equity payback (before tax).   Sum of before-tax cash flows divided by the original cash equity investment.

Equivalent uniform annual cost.   The annuity (equal annual amount) that is equivalent to some present or future worth amount, or to a cash flow consisting of various amounts at different points in time.

Equivalent uniform annual cost method.   A method for economic analysis and comparison of building projects that uses the equivalent uniform annual cost as the prime measure of performance.

Escalation.   Increase in prices; also, increase in prices for specific products over and above the general price increase due to inflation.

Escalation rate; differential escalation rate; variable escalating rate.   The rate at which specific products or services rise in cost over and above general inflation; sometimes used interchangeably with inflation rate.

Estimate. See *budget, project; contractor's estimate; construction cost; owner's estimate/architect's estimate.*

Estimate (contractor's). See *contractor's estimate.*

Expectation value. For the situation in which a course of action (alternative) will have different performance measure outcomes, depending on which of several possible states-of-the-world actually occurs: the sum of the outcomes, each multiplied by its probability of occurrence. The basis of the decision rule for such situations is to choose that alternative which has the best expectation value (i.e., the lowest expectation cost) or the highest expectation benefit.

Extended cost. The cost of a type of item (e.g., the amount of concrete needed to build a foundation wall) obtained from an estimate of the quantity takeoff of units (in this case, number of cubic feet of concrete) multiplied by the unit price (dollars per cubic foot)

Exterior surface-to-volume ratio. The ratio of exterior building surfaces (through which heat may be gained or lost—the exterior walls and roof surface) over the total volume of the building. It is a useful interim partial indicator of a building's probable energy efficiency.

Exterior-wall-to-gross-floor-area ratio (also *perimeter–area ratio*). The ratio of the exterior wall area to the total (gross) floor area.

Fast-track. A method of construction project delivery and management in which construction is started as soon as the construction documents for the first building parts (e.g., foundations) are completed, while work on the construction documents for the remainder of the building continues. This results in a shortening of the overall planning and construction time (as opposed to conventional methods where the entire set of construction documents is completed and put out for bid before construction can begin).

Feasibility analysis. Analysis of the economic/financial (but also technical, social, political) feasibility of a project. Economic feasibility analysis asks (a) whether the required rental rate of a project with a given cost and assumed economic performance will be competitive on the market; or (b) whether the affordable construction budget of a project with given competitive rental rates and assumed performance will be sufficient to get it built at market construction prices.

Financial management rate of return (FMRR). See *adjusted internal rate of return (AIRR).*

First cost. The sum of the design, construction, and construction financing cost necessary to provide a finished building or building component ready for use. Sometimes this is called the initial investment cost.

Fixed expenses. Insurance and property taxes.

Floor area ratio (FAR) (floor space index; plot ratio). A building development density measure composed of the total floor area over the area of the site: TFA/TSA.

Freehold (as opposed to leasehold tenancy). Legal ownership relationship of a person or company to a real estate property. The tenancy in this case in indefinite until it is sold and the freehold tenancy thereby transferred to a new owner.

Front door approach (financial feasibility analysis). A form of financial feasibility analysis that uses the expected capital budget (initial project cost) to determine the revenue needed to achieve the expected performance (rate of return). A project is deemed feasible if the resulting rental rate is lower than (i.e., competitive with) prevailing market rental rates for similar buildings.

Future value; future worth. The value of a benefit or a cost at some point in the future, considering the time value of money.

Function. The purpose or process served by a building, building component, or space.

Function analysis. A form of analysis that identifies and names functions of buildings, spaces, and building components, and assigns cost and worth to each function, assessing its value.

Functionally required occasion. An occasion that constitutes part of the reason for having a building; an occasion essential to the operation and use of a building.

Global optimum. The highest (best) solution in a given solution space, as opposed to the local optimum, which is a peak in performance in some region of the solution space but not the best overall.

Gradient factor (arithmetic gradient conversion factor). Factor to find the annuity A for a given year and a payment series of n years that increases uniformly year by year by an amount G. The gradient factor is $[1/i - (n/((1 + i)^{n-1}))]$.

Graduated payment mortgage. A mortgage designed to have low payments initially, rising later to correspond to expected earning patterns of a young family.

Gross floor area; total floor area. The sum of all floor areas measured flat on plan to the outside face of perimeter walls, without deduction for any openings, walls, partitions, or columns. It includes balconies, mezzanine floors where these occur within the exterior walls of the building, crawl spaces, tunnels, trenches, and floor areas with headroom of 6 feet or more, as well as penthouses, machine rooms, enclosed connections links, rooms below grade, and sidewalks, and areas of projecting columns, dormers, bay windows, and so on, provided they extend vertically for a full floor height. It excludes balconies projecting beyond exterior wall, tunnels, trenches, and so on with headroom of less than 6 feet, as well as unenclosed canopies, porches, walkways, unenclosed exterior staircases and fire escapes, exterior steps, patios, terraces, open courtyards and light wells, roof overhangs, cornices, and unfinished roof and attic areas.

Gross lease. A lease arrangement in which the lease or rent payments cover all costs of ownership except utilities (which are paid by tenant).

Growing equity mortgage. A mortgage arrangement with fixed interest rates but a schedule of rising payments to allow for faster term amortization.

Hard costs *(as opposed to soft costs).* Those project costs that must be depreciated over the life of the building, according to the tax laws.

Hurwicz alpha. A "partial optimism/pessimism" coefficient between 0 and 100% expressing where a person falls between the extremes of complete pessimism (0%) and complete (100%) optimism, used to construct a decision rule for uncertainty situations.

Iconic model. A model (representation of some system or part of reality) that visually looks like the thing represented.

Incidental occasion. An occasion that is not functionally required but occurs in between other functional requirement occasions.

Income capitalization. The process of establishing the value of a building project from its anticipated stream of income over the years.

Incremental analysis. A form of analysis that evaluates the gain in performance achieved for the increment in cost (investment) between one investment alternative and the next.

Indicator. A variable that usually changes in the same direction with another variable. Specifically: a performance indicator, changing directly with corresponding performance measures.

Inflation. A rise in the general price, usually expressed as an annual percentage rate.

Inflation rate. The rate at which prices rise annually, due to general inflation.

In-place prices. Prices of construction materials or components including their cost of transportation to the site, labor, and overhead and profit for assembly and construction (i.e., fully installed in their final place in the building).

Intangibles. Nonmonetary or generally difficult to measure costs or benefits that cannot be "touched," as opposed to tangible monetary costs or benefits.

Interest. The use fee for money lent or borrowed/invested.

Interest amount. The monetary amount of interest paid, for example, on a loan.

Interest rate. The annual percentage of, for example, a loan charged as interest.

Interim financing. Short-term financing of a building project during the construction period (construction financing) as opposed to permanent financing. Although interest is paid every month, the full sum of the principal is due upon completion of the project and then is paid with the loan of the permanent-financing mortgage.

Internal rate of return (IRR). The discount rate that makes the initial cost of the investment equal to the present value of the reversion and the present worths of annual cash flows over a given period; the compound rate of interest that, when used to discount study period costs and benefits of a project, will make the two (costs and benefits) equal.

Leasehold; leasehold tenancy (as opposed to freehold). The right to use a property that is granted to a tenant by the freehold tenant (owner) for a limited period only in return for lease or rent payments.

Leverage. The use of borrowed money to increase the return on equity investment.

Life cycle. See *study period.*

Life cycle cost analysis; life cycle cost (LCC) method. A technique of economic evaluation that sums over a given study period the costs of initial investment (less resale value), replacements, operations (including energy use), and maintenance and repair of an investment decision (expressed in present or annual value terms).

Life cycle economic performance. As opposed to life cycle cost only, a form of economic assessment that analyzes economic performance including benefits over a specified life cycle or study period.

Loan-to-value ratio. Mortgage loan divided by total project cost. A ratio or percentage specifying a lender's policy as to what fraction of a project's cost (or value, in the case of projects already completed) the lending institution is willing to finance, with the remainder to be borne by the investor as his or her equity contribution.

Locality adjustment. Adjustment of construction unit prices (e.g., in cost data manuals) and resulting estimates according to a specific geographic locality (city), based on experiential data as to how the prices in question vary in different localities.

Maintenance and repair cost. The total of labor, material, and other related costs incurred in conducting corrective and preventive maintenance and repair on a building, or on its systems and components, or on both.

Marginal utility. The increase in utility of the last added unit of some good or commodity.

MaxMin; MinMax. "Minimize your maximum losses," and "Maximize your minimum gains," respectively. A "pessimistic" decision-making rule to be used in situations of uncertainty, urging the decision-maker to select the alternative that has the least adverse consequences (i.e., the lowest loss) or the lowest cost, or the alternative that has the highest of the minimum gains over the possible states of the world.

Modified internal rate of return (MIRR). The discount rate that makes the initial cost of the investment equal to the present value of the reversion and the annual cash flows, which are reinvested at a specific rate until the end of the holding period. Synonym: adjusted internal rate of return (AIRR).

Mortgage. Loan, usually long-term (i.e., 20 or 30 years), for financing a construction project, which is repaid (amortized) with equal installments over the mortgage term.

Mortgage balance. The amount of principal still owed to the lender at a given point in time.

Mortgage constant (Capital Recovery Factor). Annual debt service divided by the original principal mortgage.

Mortgage term. The period from the commencement date of the mortgage until its maturity date.

Model. A representation of (some aspect of) reality or of a system.

Negative amortization. The effect of mortgage arrangements in which the monthly payment is not sufficient to fully cover the interest due each period, so that the amount owed actually increases instead of being amortized. The arrangement usually provides for such negative amortization for a limited period (e.g., until a young family increases its income and can afford higher mortgage payments). It is justified by the expectation of a substantial increase in the value of the property over the same time (so that the loan-to-value ratio remains the same).

Net Assignable Area (Net Usable Area or *Net Floor Area).* See *Net floor area.*

Net benefit (savings). The difference between the benefits and the costs, where both are discounted to present- or annual-value dollars.

Net floor area. The sum of all usable floor spaces measured flat on plan to the inside of enclosing walls or to lines of functional space separations. It includes areas that may be assigned to or usable for a specific function or activity. It excludes circulation and general service areas such as corridors, lobbies, toilets, restrooms, and janitors' closets, as well as mechanical, electrical, and elevator service rooms and general storage areas. It often is the same as net leasable area (NLA), which is the area that can be assigned to the exclusive use of a tenant and is the basis for calculating the rent or the lease.

Net income multiplier. Estimated project value divided by net operating income; a ratio expressing the investor's expected (desired) economic performance level.

Net operating income. Annual revenue (gross rental income) minus vacancy loss and expenses.

Net lease. A lease arrangement in which the lease or rent is calculated to cover all costs of ownership except utilities and property taxes, which are assumed to be paid directly by the tenant.

Net leasable area. The floor area that can be assigned to the exclusive use of one or more tenant(s) and is the basis for calculating the rent or the lease. It is measured from the inside of corridor and exterior walls, and the inside of fixed fire-rated walls of fire stairs and other service areas for common use. It excludes service areas, common circulation areas, elevators, ducts, and toilets. Usually this is equivalent to the net usable area, except when an entire floor is leased by one tenant, in which case the circulation area and toilets are included in the leasable area. See *Net Leasable Area* and *Net floor area.*

Net-to-Gross Ratio (efficiency ratio). Net floor area (usually net leasable area) divided by total (gross) floor area: NLA/TFA.

Network planning. Techniques such as the critical path method (CPM) or the program review evaluation technique (PERT) used to plan and maintain control of schedules of project activities in, for example, a construction project.

Net worth of property (project net worth). Resale price minus resale costs, mortgage balance, and working capital loan.

Nominal discount rate. Rate of interest reflecting the time value of money, stemming from both inflation and the real earning power of money over time. *Note:* This is the discount rate used in discount formulae or in selecting discount factors when future benefits and costs are expressed in current dollars.

Objectification. The process of explaining how (subjective) judgments of a person depend on other judgments or on objective measures of performance.

Objective. Noun: a specific description of a goal, intention, or target. Usually the aim is to specify an objective in such a way that one can test or measure specifically whether and to what extent it has been reached. Adjective: having the property of objectivity, that is, being dependent on observable, measurable facts, not influenced by feelings.

Occasion. An event, situation, or process involving human users, to be housed in a building.

Occasion opportunity. The potential for an occasion to be performed or to occur in a space or building.

Occasion opportunity density. The number of occasion opportunities per unit of floor area.

Operating cost. The expenses incurred during the normal operation of a building or a building system or component, including labor, materials, utilities, and other related costs.

Operating expense ratio. Total operating expenses divided by potential gross income.

Opportunity cost of capital. The rate of return available on the next best available investment of comparable risk. The term is used interchangeably with discount rate, understood as an investor's average expected rate of return on his or her other investments.

Optimal. Being the best.

Optimization. The process of finding the best solution.

Optimum. The best.

Overall rate. Rate of return of an investment (in a loan, the rate at which the principal is paid back, which is a function of the duration or term of the loan) plus rate of return on the investment (the interest rate on the loan).

Owner's estimate/architect's estimate. Any estimate of initial costs of a building project that looks at the combined costs of materials, labor, overhead and profit, and so on, involved in installing the respective building components in place (as opposed to the contractor's estimate, where the costs of materials, labor, and overhead are kept separate; the contractor can manipulate only the overhead part of an estimate in the effort to make a bid competitive).

Payback method. A technique of economic evaluation that determines the time required for the cumulative benefits from an investment to recover the investment cost and other accrued costs. See *Discounted payback period* and *simple payback period.*

Performance indicator. A variable that indicates or hints at how well a given design is likely to perform (as it eventually would be measured by a performance variable), or that describes the contribution the design solution itself makes toward a good performance (ignoring the influence of other context variables not under the designer's control but that also determine performance).

Performance measure. A variable that measures how well a design solution meets the client's objectives or serves its purpose. It should allow the designer to tell which of two solutions is "better" than the other.

Performance variable.   A variable describing a project that measures how well the project serves the client's/owner's interests.

Perimeter–area ratio.   A performance indicator composed of the length of a (typical) floor plan's perimeter (exterior walls) over the enclosed floor area.

Permanent financing.   Financing of a building project (mortgage) in which the debt service payments are spread out over a long period, typically 20 or 30 years.

Population density.   A population density measure expressed as number of people per unit of land; for example, persons per square mile, persons per acre, households per square mile, households per acre. Also, the number of dwelling units per unit of land is used.

Present value.   "Today's" worth of future benefits discounted at a given rate of return. The value of a benefit or cost found by discounting future cash flow to the base time (present worth).

Present value factor; present worth factor.   The discount factor used to convert future values (benefits and costs) to present values.

Present worth.   See *Present value.*

Principal.   The mortgage debt; the original sum of a loan; the portion of a mortgage payment that consists of repayment of the principal as opposed to interest on the outstanding principal balance.

Private goods.   Goods that allow the purchaser to exclude others from enjoying their use or consumption.

Pro forma analysis (Real Estate Pro Forma Analysis; Real Estate Pro Forma Statement).   A form of real estate financial feasibility analysis that looks at all essential components of the financing package for a project to determine whether the project will achieve the desired performance, and whether the required equity funds are on hand to get the project started.

Project value.   As opposed to project cost (the cost of purchase or construction), the capitalized value of the expected stream of income over the years.

Property cap rate; property capitalization rate.   The rate at which a property earns money—the inverse of the net income multiplier (NIM).

Quantity survey; quantity takeoff.   An item-by-item count or inventory of all materials and building components needed in a building project.

Quantity survey–based estimating method.   A method for estimating initial cost of a building project that consists of establishing estimated prices for each item, component, or material in a building and multiplying these prices by the number of items or amounts of materials needed.

Quasi-collective goods.   Goods bought and sold as if they were private goods (e.g., education), but from which there accrue general benefits to society that justify subsidies for such goods.

Rate of return (return rate).   A ratio of returns earned from an investment or project over the (initial) cost or equity contribution. Different types of rate of return are distinguished, depending on whether the returns are defined as before or after subtracting debt service or be-

fore or after taxes, whether returns are discounted to account for the time value of money, whether the net returns are assumed to be reinvested and earning interest or not, and whether the investment is total initial cost or equity contribution.

Rate of return on total capital investment (overall rate).   Net operating income divided by initial project cost.

Real discount rate.   The rate of interest reflecting that portion of the time value of money related to the real earning power of money over time. *Note:* This is the rate used in discount factors when future benefits and costs are expressed in constant dollars (as opposed to the nominal discount rate).

Real dollars.   See *constant dollars*

Recapture.   Process by which allowable excess depreciation is taxed when a property is sold.

Refinancing surplus.   Excess mortgage proceeds resulting from the commencement of a new mortgage on the property.

Regret.   A measure of the difference between the benefits or payoffs received from a selected alternative given a state-of-the world (context situation) and the payoff that could have been received from the best alternative given the same state-of-the-world.

Replacement cost.   Building component replacement and related costs, included in the capital budget, that are expected to be incurred during the study period.

Replacement reserve.   Funds set aside to cover future projected renovations and replacement expenses.

Resale value (sale value).   The monetary sum expected from the disposal of an asset at the end of its economic life, its useful life, or the study period.

Revenues.   Potential gross income. Rent/square foot times gross leasable area.

Reversion.   Benefits received at the end of the investment holding period, usually resulting from selling the property.

Salvage value.   The value of an asset (usually assigned for tax computation purposes) that is expected to remain at the end of the depreciation period. Also, scrap value—the amount of money that, for example, a tool or a machine can be sold for at disposal that reflects only the value of reusable material (e.g., iron), not the value of the tool as such.

Savings-to-investment ratio (SIR).   Either the ratio of present value savings to present value investment costs or the ratio of annual value savings to annual value of investments costs. This refers to a comparison between two alternative choice for a building project or special components (e.g., different fuel systems for the HVAC system of a building); usually, one alternative is the conventional state-of-the-art solution serving as the baseline alternative, the other being an innovative, energy-saving alternative that would require a higher initial investment.

Sensitivity analysis.   A test of the outcome of an analysis by altering one or more parameters from an initially assumed value(s).

Service area.   The area consisting of circulation, lobby, stairs, toilets, mechanical and duct rooms, and so on, that is not part of the usable and therefore rentable area of a building.

Shared-appreciation mortgage.   A mortgage arrangement that exploits the ability of large companies (developer, builder) to secure lower interest loans than an individual home buyer. In return for below-market interest rates, the buyer agrees to share a portion of the sale price with the lender when the property is sold before the mortgage term. The expectation is that the property will appreciate substantially.

Simple payback method.   Method for evaluating building project alternatives using the payback period as the measure of performance where, however, the benefit cash flows have not been discounted to present worth.

Simple payback period.   The time required for the cumulative benefits from an investment to pay back the investment cost and other accrued costs, not considering the time value of money (not discounting benefits or costs to present worth).

Sinking fund factor.   Conversion factor used to convert a future worth amount into its annuity series equivalent: $(A/F,i,n)$ or $i/(1 + i)^n$.

Soft costs (as opposed to hard costs).   Those parts of the initial project costs that according to the tax laws must be expensed initially and cannot be depreciated over the life of the building.

Solution space.   The range of possible solutions for a problem, within a system of constraints.

Spendable cash after taxes (after-tax cash flow).   Revenue minus vacancy loss, expenses, debt service, income taxes, and tax sheltered cash, if any.

Study period (life cycle, time horizon).   The length of time over which an investment is analyzed.

Subjective.   Dependent on or reflecting personal feelings, preferences, opinion, as opposed to objective facts and measurements.

Sunk cost.   A cost that already has been incurred and will not be affected by an investment decision (should not be considered in making a current investment decision).

Surface–volume ratio.   The ratio formed by the surface area of a building (exterior skin) to its volume.

Symbol.   A short, easy-to-remember (mnemonic) name for a variable or an entity; some form, shape, or concept that stands for another concept or reality.

Symbolic model.   A model (i.e., representation of some system) in which the modeled reality is described by means of symbols (standing, e.g., for variables).

System.   (1) A set of objects together with relationships between the objects and their attributes or properties (general systems science definition). (2) A group of building components serving a major common function; for example, the structural system, or the exterior closure system. (*see* systems method, systems estimate). (3) A set of preestablished, coordinated building components and procedures designed for quick (often dry) assembly of the same elements in different configurations, and/or in view of later easy repositioning and reassembly without wasteful demolition, cutting, or other adaptation that would make reuse impossible.

Systems estimate.   A cost estimate for a building project based on a breakdown of the structure into component systems (e.g., exterior closure or structural system) for which separate estimates are made (these estimates could be based on data from similar systems, or on detailed unit price estimates). Also, an estimate of initial building cost resulting from the systems method of estimating.

Systems method.   Method for estimating initial cost of building projects using an overall construction price ($/sf of TFA) and estimated percentages of cost for the various systems and subsystems.

Tax shelter cash.   Cash credit resulting from a negative income tax liability.

Taxable income.   Net operating income minus interest and depreciation.

Teleological.   Goal-seeking; purposeful (e.g., teleological systems; systems that pursue a goal or serve a purpose.)

Time horizon.   See *study period*. The length of time or planning horizon considered in an economic performance analysis during which incurred costs or benefits received are of concern to the investor.

Time series.   A set of data observations of some variable over time; used in forecasting as time series extrapolation by extending the trend observed in the past into the future.

Time value of money.   The time-dependent value of money, stemming both from changes in the purchasing power of money (i.e., inflation or deflation) and from the real earning potential of alternative investments over time, as well as from the fact that people assign a higher value to money received "now" as opposed to the same amount received at some future date.

Total floor area (TFA) (gross floor area).   The area of all floors of a building, measured to the outside of walls and construction, according to, for example, AIA conventions.

Total investment value.   The sum of the present value of spendable cash after taxes, present value of the reversion, and the mortgage loan.

Trade breakdown method.   Method for estimating initial cost of building projects based on the percentages of the work carried out by the different trades involved in its construction (e.g., masonry, concrete work, carpentry, painting, etc.).

Triple net lease.   A lease arrangement in which the rent or lease payments cover only debt service, owner's overhead and profit, and sales taxes on rent, if any. The tenant is responsible for all other expenses, including utilities, real estate taxes, insurance, and maintenance.

Unit-of-use method.   Method of estimating initial cost of building projects based on functional use units and their unit prices. Examples include student stations for schools, patient beds for hospitals, boarding gates for airports, and so on.

Unit price.   The price (often used interchangeably with *unit cost*) for one unit of something; for example, dollars per square foot of roof covering or dollars per metal wall stud.

Unit price estimate.   Estimates prepared by means of the unit price method or *quantity takeoff–based method*.

Useful life.   The period of time over which an investment is considered to meet its original objective.

Vacancy rate.   That percentage of the leasable or rentable space of a building that will be vacant at any given time (on the average) because of tenants moving and the time lapse between tenants, as well as the time needed to find new tenants. It is the inverse of occupancy rate: vacancy rate = (1 − occupancy rate). It serves as a measure of attractiveness or competitiveness of the facility, and is used to establish the actual (effective) rental income from a project. Usually it is lumped with bad debt losses—bounced checks, tenants who are late paying their rent or move out without paying their last month's rent, and so forth.

Value engineering.   A systematic approach used to identify the function of buildings or building components, to assess their worth and cost, and to look for ways in which the purpose (function) of the component can be served at lower cost.

Variable.   Characterization of something that can take on any one of a number of different values on a scale (i.e., the values can vary—e.g., the length of some item).

Variable expenses.   Expenses whose magnitude is not fixed (e.g., debt service or real estate taxes) but depends on the activities and policies of the user and owner (e.g., utilities, maintenance, management, repair, redecoration, grounds keeping, and trash removal).

Volume.   The enclosed cubic content, measured to the outside faces of perimeter walls and from the underside of the lowest floor construction system to the average height of the surface of the finished roof, including all voids within these limitations. It includes all spaces in the gross floor area calculation; chimneys; stacks or other large features projecting above the roof level; bay and dormer windows and other large features projecting beyond the perimeter wall facade; and pits, trenches, and depressions projecting below the lowest floor construction. It excludes all spaces excluded from the gross floor area calculation, such as enclosed skylights less than 6 sf, smoke hatches, fan housings, and so on, as well as interior open courts and light wells.

Volume method.   Method for estimating initial cost of building projects based on volume units (e.g., cubic feet or cubic meters) and unit prices expressed as price per volume unit (e.g., $/cubic foot). For example, building cost = no. of cubic feet times price per cubic foot.

Whole unit method.   Method for estimating initial cost of building projects based on the estimated price per standard unit (e.g., of a typical house), in projects involving a number of such units.

Working capital loan.   Short-term financing that provides cash to cover the losses created by a negative cash flow from operations.

Yield (after-tax cash on cash rate of return, or yield rate).   A return rate formed by spendable cash after taxes (SCAT) over initial equity investment: SCAT/CEQC.

Appendix 15
References

AIA (The American Institute of Architects). 1977. *Life Cycle Cost Analysis—A Guide For Architects*. Washington, DC: The American Institute of Architects.

AIA (The American Institute of Architects). 1978. *Life Cycle Cost Analysis—Using it in Practice*. Washington, DC: The American Institute of Architects.

Alexander, Christopher et al. 1977. *A Pattern Language*. Oxford University Press.

ASTM (The American Society for Testing and Materials). Standards documents, reprinted from *The Annual Book of ASTM Standards:*

- E 1074-85: "Standard Practice for Measuring Internal Rates of Return for Investment in Buildings and Building Systems."
- E 1074-85: "Standard Practice for Measuring Net Benefits for Investments in Buildings and Building Systems."
- E 833-85a: "Standard Definitions of Terms Relating to Building Economics."
- E 1185-87: "Standard Guide for Selecting Economic Methods for Evaluating Investments in Buildings and Building Systems."

Bechtel, Robert B., R.W. Marans, and W. Michelson, eds. 1987. *Methods in Environmental and Behavioral Research*. New York: Van Nostrand Reinhold.

Bon, Ranko. 1989. *Building as an Economic Process*. Englewood Cliffs, NJ: Prentice-Hall.

Canestaro, James. 1979. *Real Estate Financial Feasibility Handbook*. Madison, WI: Department of Real Estate and Urban Land Economics, University of Wisconsin.

Churchman, C. West. 1971. *The Design of Inquiring Systems*. New York: Basic Books.

Churchman, C. West. 1979. *The Systems Approach*. Revised edition. New York: Laurel Books.

Class, Robert Allan, and R.E. Koehler, eds. 1976. *Current Techniques in Architectural Practice*. Washington, DC: The American Institute of Architects.

Collier, Courtland A. and Don A. Halperin. 1985. *Construction Funding*. Second edition. New York: John Wiley & Sons.

Cox, Billy J. and F. William Horsley. 1983. *Means Square Foot Estimating*. Kingston. MA: R.S. Means Company.

Dell'Isola, Alphonse J. and Stephen J. Kirk. 1981. *Life Cycle Costing for Design Professionals*. New York: McGraw-Hill.

Dell'Isola, Alphonse J. and Stephen J. Kirk. 1983. *Life Cycle Cost Data*. New York: McGraw-Hill.

Dorfman, R., ed. 1965. *Measuring Benefits of Government Investments*. Washington, DC: Brookings Institute.

Goldthwaite, Richard. 1980. *The Building of Renaissance Florence*. Baltimore, MD: The Johns Hopkins University Press.

Hall, and Fagen. 1956. "Definition of System." in: *Yearbook of the Society for General System Research*. Vol. 1. Ann Arbor, MI: Society for the Advancement of General Systems Research.

Helmer, Olaf. 1968. *Social Technology*. New York: Basic Books.

Johnson, Robert E. 1990. *The Economics of Building*, New York: John Wiley & Sons.

Kemper, Alfred. 1979. *Architectural Handbook*. New York: John Wiley & Sons.

Killingsworth, Roger. 1988. *Cost Control in Building Design*. Kingston, MA: R. S. Means Co.

Kirk, Stephen J. and Kent F. Spreckelmeyer. 1988. *Creative Design Decisions*. New York: Van Nostrand Reinhold.

Mann, Thorbjoern. 1977. "*Argument Assessment for Design Decisions*," Dissertation submitted in partial fulfillment of requirements for the degree of Doctor of Philosophy, University of California, Berkeley.

Mann, Thorbjoern. 1980a. "Some Limitations of the Argumentative Model of Design," *Design Methods and Theories*, Vol. 14, No. 1.

Mann, Thorbjoern. 1980b. "Places and Occasions," *Design Methods and Theories*, Vol. 14, No. 2.

Mann, Thorbjoern. 1986. "Procedural Building Blocks," in Robert Trappl, ed *Proceedings, Eighth European Meeting for Cybernetics and Systems Research*, Vienna, Austria.

Mann, Thorbjoern. 1990. "The Need for Intermediate Level Paradigms." Robert Trappl, ed. *Proceedings, Tenth European Meeting for Cybernetics and Systems Research*, Vienna, Austria.

Meadows, Dennis et al. 1972. *The Limits of Growth*. Washington, DC: Potomac Associates.

R.S. Means Co. *Means Square Foot Costs*. Annual Construction Cost Guide. Kingston, MA.

R.S. Means Co. *Means Building Construction Cost Data*. Annual Guide. Kingston, MA.

R.S. Means Co. *Means Facilities Cost Data*. Annual Cost Guide. Kingston, MA.

Mishan, E. J. 1976. *Cost-Benefit Analysis*. New expanded edition. New York: Praeger.

Musso, Arne and H. W. Rittel. 1968. "Über das Messen der Güte von Gebäuden," in *Arbeitsberichte zur Planungsmethodik 1*. Stuttgart, Germany: Institut für Grundlagen der Modernen Architektur, Stuttgart University.

Newman, Oscar. 1972. *Defensible Space*. New York: Macmillan.

Newnan, Donald G. 1988. *Engineering Economic Analysis*. Third ed. San Jose, Engineering Press.

Peña, William, with Steven Marshall and Kevin Kelley. 1990. *Problem Seeking, An Architectural Programming Primer*. Revised edition. Washington, DC: The AIA Press.

Riggs, James and Thomas M. West. 1986. *Engineering Economics*. Third edition. New York: McGraw-Hill.

Ring, Alfred, and Jerome Dasso. 1985. *Real Estate, Principles and Practice*. Tenth edition. Englewood Cliffs, NJ: Prentice-Hall.

Rittel, Horst W. 1970. "Der Planungsprozess als Iterativer Vorgang von Varietätserzeugung und Varietätseinschränkung," in: *Arbeitsberichte zur Planungsmethodik 4*. Stuttgart, Germany: Institut für Grundlagen der Modernen Architektur, Stuttgart University.

Rittel, Horst and M. Webber. 1973. "Dilemmas in A General Theory of Planning" in: *Policy Sciences* No. 4, pp. 155–169.

Ruegg, Rosalie T. and H. E. Marshall. 1990. *Building Economics, Theory and Practice*. New York: Van Nostrand Reinhold.

Sassone, Peter G. and William A. Schaffer. 1979. *Cost-Benefit Analysis*. New York: Academic Press.

Starfield, Anthony M., Karl A. Smith, and Andrew L. Bleloch. 1990. *How to Model It, Problem-Solving for the Computer Age*. New York: McGraw-Hill.

Stone, Peter A. 1980. *Building Design Evaluation*. Third edition. London and New York: Spon.

Stone, Peter A. 1986. *Building Economy: Design, Production, and Organization*. Third edition. Oxford, U.K.: Pergamon Press.

Swinburne, Herbert. 1980. *Design Cost Analysis for Architects and Engineers*. New York: McGraw-Hill.

U.S. Government. 1973. *Benefit-Cost Analysis of Federal Programs*. Washington, DC: U.S. Government Printing Office.

Ward, Robertson. 1987. "Office Building Systems Performance and Functional Use Costs." *CIB Proceedings of the Fourth International Symposium on Building Economics*. Copenhagen, Denmark: Danish Building Research Institute.

Index